T0269610

CAMBRIDGE LIBRARY COLLECTION

Books of enduring scholarly value

Astronomy

From ancient times, humans have tried to understand the workings of the world around them. The roots of modern physical science go back to the very earliest mechanical devices such as levers and rollers, the mixing of paints and dyes, and the importance of the heavenly bodies in early religious observance and navigation. The physical sciences as we know them today began to emerge as independent academic subjects during the early modern period, in the work of Newton and other 'natural philosophers', and numerous sub-disciplines developed during the centuries that followed. This part of the Cambridge Library Collection is devoted to landmark publications in this area which will be of interest to historians of science concerned with individual scientists, particular discoveries, and advances in scientific method, or with the establishment and development of scientific institutions around the world.

The Description and Use of the Globes, and the Orrery

Born in rural Wales, to which he always felt a close connection, Joseph Harris (c.1704–64) moved to London in 1724, presenting the Astronomer Royal, Edmond Halley, with a testimonial of his mathematical ability. Harris then found work as an astronomer and teacher of navigation; his observations of magnetism and solar eclipses taken in Vera Cruz in 1726 and 1727 were relayed to the Royal Society by Halley. Harris' illustrated introduction to the solar system was originally printed for the instrument-maker Thomas Wright and the globe-maker Richard Cushee; it is here reissued in its 1731 first edition. Clearly describing the use of astronomical apparatus such as globes and orreries, it proved very popular, going through fourteen printings by 1793. Harris starts with an overview of the solar system and the fixed stars, and then shows how to solve astronomical problems using globes and orreries.

Cambridge University Press has long been a pioneer in the reissuing of out-of-print titles from its own backlist, producing digital reprints of books that are still sought after by scholars and students but could not be reprinted economically using traditional technology. The Cambridge Library Collection extends this activity to a wider range of books which are still of importance to researchers and professionals, either for the source material they contain, or as landmarks in the history of their academic discipline.

Drawing from the world-renowned collections in the Cambridge University Library and other partner libraries, and guided by the advice of experts in each subject area, Cambridge University Press is using state-of-the-art scanning machines in its own Printing House to capture the content of each book selected for inclusion. The files are processed to give a consistently clear, crisp image, and the books finished to the high quality standard for which the Press is recognised around the world. The latest print-on-demand technology ensures that the books will remain available indefinitely, and that orders for single or multiple copies can quickly be supplied.

The Cambridge Library Collection brings back to life books of enduring scholarly value (including out-of-copyright works originally issued by other publishers) across a wide range of disciplines in the humanities and social sciences and in science and technology.

The Description and Use of the Globes, and the Orrery

To which is Prefixed, by way of Introduction,
a Brief Account of the Solar System

JOSEPH HARRIS

CAMBRIDGE
UNIVERSITY PRESS

CAMBRIDGE
UNIVERSITY PRESS

University Printing House, Cambridge, CB2 8BS, United Kingdom

Cambridge University Press is part of the University of Cambridge.
It furthers the University's mission by disseminating knowledge in the pursuit of
education, learning and research at the highest international levels of excellence.

www.cambridge.org
Information on this title: www.cambridge.org/9781108080187

This edition first published 1731
This digitally printed version 2015

ISBN 978-1-108-08018-7 Paperback

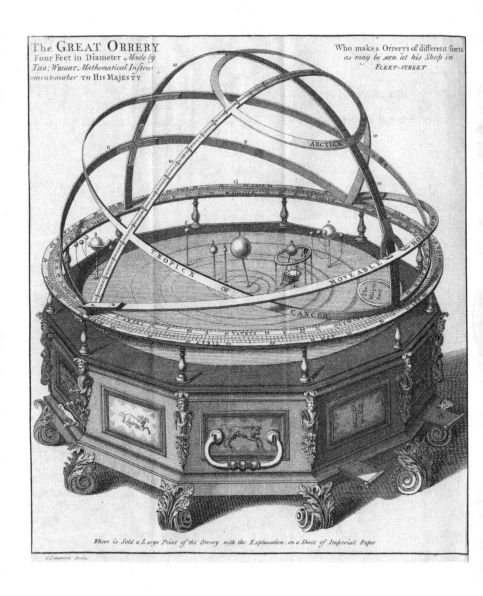

The GREAT ORRERY
Four Feet in Diameter *Made by*
Tho: Wright, Mathematical Instru-
=ment-maker TO HIS MAJESTY

Who makes Orrerys of different sorts
as may be seen at his Shop in
FLEET-STREET

Where is Sold a Large Print of the Orrery with the Explanation on a Sheet of Imperial Paper

The material originally positioned here is too large for reproduction in this
reissue. A PDF can be downloaded from the web address given on page iv
of this book, by clicking on 'Resources Available'.

THE
Description and *Use*
OF THE
GLOBES,
AND THE
ORRERY.

To which is prefixed,

By way of INTRODUCTION, a Brief Account of the SOLAR SYSTEM.

By *J. HARRIS.*

LONDON:
Printed for THOMAS WRIGHT, Mathematical Inftrument-Maker, at the *Orrery and Globe* near *Salisbury-Court*; and RICHARD CUSHEE, Globe-Maker, at the *Globe and Sun* between St. *Dunftan's* Church and *Chancery-Lane*; both in *Fleet-ftreet.* M.DCC.XXXI.

THE

CONTENTS.

The INTRODUCTION: *Containing a brief Account of the Solar Syftem, and of the fixed Stars.*

A 2 *The*

The DESCRIPTION *and* USE *of the* CELESTIAL *and* TERRESTRIAL GLOBES. 35

PROB.

The CONTENTS: v

PROB.

vi   The CONTENTS.

The CONTENTS. vii

The DESCRIPTION of the ORRERY. 152

THE

INTRODUCTION,

CONTAINING

A Brief Account of the SOLAR SYSTEM, and of the FIXED STARS.

SECT. I.

Of the Order, and Periods of the Primary Planets revolving about the Sun ; and of the Secondary Planets round their respective Primaries.

THE Sun is placed in the midst of an immense Space, wherein Six Opaque Spherical Bodies revolve *Planets.* about him as their Center. These wandring Globes are called the *Planets,* who at different Distances, and in different

B   Periods,

Periods, perform their Revolutions from West to East, in the following Order.

1. ☿ *Mercury* is nearest to the Sun of all the Planets, and performs its Course in about three Months. 2. ♀ *Venus* in about seven Months and a half. 3. ⊕ The *Earth* in a Year. 4. ♂ *Mars* in about two Years. 5. ♃ *Jupiter* in twelve. And lastly, ♄ *Saturn*, whose * *Orbit* includes all the rest, spends almost 30 Years in one Revolution round the Sun.   The distances of the Planets from the Sun are nearly in the same proportion, as they are represented in *Plate* 1. *viz.* Supposing the Distance of the Earth from the Sun to be divided into 10 equal parts; that of *Mercury* will be about 4 of these parts; of *Venus* 7; of *Mars* 15; of *Jupiter* 52; and that of *Saturn* 95.

The Orbits of the Planets are not all in the same Plane, but variously inclined to one another; so that supposing one of them to coincide with the above Scheme, the others will have one half above, and the other half below it; intersecting one another in a Line passing through the Sun. The Plane of the Earth's Orbit, is called the

---

The Characters placed before the Names of the Planets, are for Brevity's sake commonly made use of by Astronomers, instead of the Words at length, as ♀ for *Venus*, &c.

* By the Orbit of a Planet, is commonly understood the Tract or Ring described by its Center round the Sun; but by the Plane of the Orbit is meant a flat Surface extended every way thro' the Orbit infinitely.

Plate 1.

Page 1

THE SOLAR SYSTEM Or the Orbits of the Planets

according to their mean ♄ *distances from the Sun*

♃

♂

Saturn

Jupiter

Magnitudes

● Mars
◑ the Earth
● Venus
• Mercury
• the Moon

the Sun 10 Inches Diameter
according to this proportion

*N.B. The Orbits of the Secondary Planets are here 50 times
greater in proportion than the distances of the Primary
Planets from the Sun*

R. Cushee sculp.

the *Ecliptick*; and this the Astronomers Ecliptick.
make the Standard, to which the Planes
of the other Orbits are judged to incline.
The right Line passing thro' the Sun, and
the common Intersection of the Plane of
the Orbit of any Planet and the Ecliptick,
is called the *Line of the Nodes* of that Nodes.
Planet; and the Points themselves, wherein
the Orbit cuts the Ecliptick, are called the
Nodes.

The Inclinations of the Orbits of the
Planets to the Plane of the Ecliptick, are
as follow, *viz.* The Orbit of *Mercury* makes
an Angle with it of almost 7 degrees;
that of *Venus* something above $3\frac{1}{3}$ degrees;
of *Mars*, a little less than 2 degrees; of *Ju-*
piter, $1\frac{1}{3}$ degrees; and of *Saturn*, about
$2\frac{1}{2}$ degrees. The Orbits of the Planets are
not Circles, but Ellipses or Ovals. What
an Ellipsis is, may be easily understood from
the following description. Imagine two
small Pegs fixed upright on any Plane, and
suppose them tyed with the ends of a Thread
somewhat longer than their distance from
one another: now if a Pin be placed in the
double of the Thread, and turned quite
round, (always stretching the Thread with
the same force) the Curve described by this
Motion is an *Ellipsis*. The two Points
where the Pegs stood, (about which the
Thread was turned) are called the *Foci* of
that Ellipsis; and if without changing the

B 2   length

The INTRODUCTION.

length of the Thread, we alter the Pofition of the Pegs, we fhall then have an Elliplis of a different kind from the former; and the nearer the *Focus's* are together, the nearer will the Curve defcribed be to a Circle; until at laft the two *Focus's* coincide, and then the Pin in the doubling of the Thread will defcribe a perfect Circle. The Orbits of all the Planets, have the Sun in one of their *Focus's*; and half the diftance between the two *Focus's* is called the *Excentricity* of the Orbits. This Excentricity is different in all the Planets; but in moft of them it is fo fmall, that in little Schemes or Inftruments, made to reprefent the Planetary Orbits, it need not be confidered.

Excentricity.

The Six Planets above-mention'd, are called *Primaries,* or *Primary Planets*; but befides thefe, there are ten other lefler Planets, which are called *Secondaries, Moons* or *Satellits.* Thefe Moons always accompany their refpective Primaries, and perform their Revolutions round them, whilft both together are alfo carried round the Sun. Of the Six Primary Planets, there are but three, as far as Obfervation can affure us, that have thefe Attendants, *viz.* the *Earth, Jupiter,* and *Saturn.*

Primary Planets.

Secondary Planets.

The Earth is attended by the *Moon,* who performs her Revolution in about $27\frac{1}{3}$ days, at the diftance of about 30 Diameters of the Earth from it; and once a Year is carried

carried round the Sun along with the Earth.

Jupiter has four *Moons* or *Satellits* ; the *first* or innermoſt performs its Revolution in about one Day and 18 ½ Hours, at the diſtance of 5 ⅓ Semidiameters of *Jupiter* from his Center ; the *ſecond* revolves about *Jupiter* in 3 Days and 13 Hours, at the diſtance of 9 of his Semidiameters ; the *third* in 7 Days and 4 Hours, at the diſtance of 14 ⅓ Semidiameters ; the *fourth* and *outermoſt*, performs its Courſe in the ſpace of 16 Days 17 Hours, and is diſtant from *Jupiter*'s Center, 25 ⅓ of his Semidiameters.

Jupiter's four Moons.

Saturn has no leſs than five *Satellits* ; the *firſt* or innermoſt revolves about him in 1 Day and 21 Hours, at the diſtance of 4 ⅜ Semidiameters of ♄ from his Center ; the *ſecond* compleats its Period in 2 ¾ Days at the diſtance of 5 ⅖ of his Semidiameters ; the *third*, in about 4 ¼ Days at the diſtance of 8 Semidiameters ; the *fourth* performs its Courſe in about 16 Days, at the diſtance of 18 Semidiameters ; the *fifth* and outermoſt takes 79 ⅓ Days to finiſh his Courſe, and is 54 Semidiameters of *Saturn*, diſtant from his Center. The Satellits as well as the Primaries perform their Revolutions from *Weſt* to *Eaſt* : the Planes of the Orbits of the Satellits of the ſame Planet, are variouſly inclined to one another, and conſequently are inclined to the Plane of the Orbit of their Primary

Saturn has five Moons.

Be-

Saturn's
Ring.

Befides thefe Attendants, *Saturn* is encompaffed with a thin plain Ring that does
no where touch his Body : the Diameter of
this Ring is to the Diameter of *Saturn* as
9 to 4; and the void fpace between the
Ring and the Body of *Saturn*, is equal to
the breadth of the Ring it felf; fo that in
fome fituations the Heavens may be feen
between the Ring and his Body. This fur-
prifing Phenomenon of *Saturn's Ring* is a
modern difcovery; neither were the Satellits of *Jupiter* and *Saturn* known to
the Ancients. The *Jovial* Planets were firft
difcovered by the famous *Italian* Philofo-
pher *Galilæus*, by a Telefcope which he
firft invented; and the celebrated *Caffini*,
the *French* King's Aftronomer, was the firft
that faw all the *Satellits* of *Saturn*; which
by reafon of their great diftances from the
Sun, and the fmallnefs of their own Bodies,
cannot be feen by us, but by the help of
very good Glaffes.

Annual
Motion.

The Motion of the primary Planets round
the Sun (as alfo of the Satellits round
their refpective Primaries) is called their
Annual Motion; becaufe they have one
Year or Alteration of Seafons complete in
one of thefe Revolutions. Befides this
Annual Motion, four of the Planets, *viz.*
Venus, the *Earth*, *Mars*, and *Jupiter*, re-

Diurnal
Motion.

volve about their own *Axes*, from *Weft*
to *Eaft*; and this is called their *Diurnal*
Mo-

Motion. For by this Rotation each point of their Surfaces, is carried fuccefsively towards or from the Sun, who always illuminates the Hemifphere which is next to him, the other remaining obfcure : and while any Place is in the Hemifphere illuminated by the Sun, it is *Day* ; but when it is carried to the obfcure Hemifphere, it becomes *Night* ; and fo continues until by this Rotation the faid place is again enlightned by the Sun.

The *Earth* performs its Revolution round its Axis in 23 Hours 56 Minutes ; *Venus* in 23 Hours ; *Mars* in 24 Hours and 40 Minutes ; and *Jupiter* moves round his own Axis in 9 Hours and 56 Minutes. The Sun alfo is found to turn round his Axis from Weft to Eaft in 27 Days : And the Moon, which is neareft to us of all the Planets, revolves about her Axis in a Month, or in the fame fpace of Time that fhe turns round the Earth ; fo that the Lunarians have but one Day throughout their Year. *Diurnal Motion of the* ⊕, ♀, ♂, *and* ♃. ⊙ *and* ☽ *likewife turn round their Axes.*

I. The Planets are all opaque Bodies, having no Light but what they borrow from the Sun : For that fide of them which is next towards the Sun, has always been obferved to be illuminated, in what pofition foever they be ; but the oppofite fide, which the Solar Rays do not reach, remain dark and obfcure ; whence it is evident that they have no light but what proceeds from the *The Planets are Opaque and Globular.*

Sun :

Sun : for if they had, all parts of them would be lucid without any darkneſs or ſhadow. The Planets are likewiſe proved to be Globular, becauſe let what part ſoever of them be turned towards the Sun, its Boundary, or the Line ſeparating that part from the oppoſite, always appears to be Circular; which could not happen if they were not Globular.

The Pla- II. That the Earth is placed betwixt the
nets turn Orbs of *Mars* and *Venus*, and that ☿, ♀,
round the ♂, ♃, and ♄, do all turn round the Sun;
Sun. is proved from Obſervations as follow :

1. Whenever *Venus* is in Conjunction with the Sun, that is, when ſhe is in the ſame Direction from the Earth, or towards the ſame part of the Heavens the Sun is in; ſhe either appears with a bright and round Face like a full Moon, or elſe diſappears; or if ſhe is viſible, ſhe appears horned like a new Moon; which Phenomena could never happen if ♀ did not turn round the Sun, and was not betwixt him and the Earth : For ſince all the Planets borrow their Light from the Sun, it is neceſſary that ♀'s lucid Face ſhould be towards the Sun; and when ſhe appears fully illuminated, ſhe ſhews the ſame Face to the Sun and Earth; and at that time ſhe muſt be above or beyond the Sun, for in no other Poſition could her illuminated Face be ſeen from the Earth. Farther, when ſhe diſappears, or,

if

if vifible, appears horned; that Face of
hers which is towards the Sun, is either
wholly turned from the Earth, or only a
fmall part of it can be feen by the Earth;
and in this cafe fhe muft of neceflity be
betwixt us and the Sun.   Let S be the *Sun*, *Plate* 2.
T the *Earth*, and V *Venus*, having the *Fig.* 1. 2.
fame Face prefented both towards the *Sun*
and *Earth :* here it is plain that the Sun is
betwixt us and *Venus*, and therefore we
muft either place *Venus* in an Orbit round
the *Sun*, and likewife betwixt him and
us, as in *Fig* 1. or elfe we muft make the
Sun to move round the Earth in an Orbit
within that of *Venus*, as in *Fig* 2.   Again,
after *Venus* difappears, or becomes horned, at her * ♂ with the ☉, fhe then muft be
betwixt us and the Sun, and muft move
either in an Orbit round the Sun, and betwixt us and him, as in *Fig.* 1. or elfe
round the Earth, and betwixt us and the
Sun, as in *Fig.* 2. But *Venus* cannot move
fometimes within the Sun's Orbit, and
fometimes without it, as we muft fuppofe
if fhe moves round the Earth ; therefore it
is plain that her Motion is round the Sun.

Befides the forcgoing, there is another
Demonftration that *Venus* turns round the
Sun, in an Orbit that is within the Earth's ;
becaufe fhe is always obferved to keep near
the

---

* ♂ is a Mark commonly ufed for Conjunction: thus ♂
with the ☉, is to be read Conjunction with the Sun.

the Sun, and in the same Quarter of the Heavens that he is in : for she never recedes from him more than about $\frac{1}{9}$ of a whole Circle; and therefore can never come in opposition to him; which would necessarily happen, did she perform her Course round the Earth either in a longer or shorter time than a Year. And this is the reason why *Venus* is never to be seen near midnight, but always either in the Morning or Evening, and at most not above three or four Hours before Sun-rising, *Why* Venus or after Sun-setting. From the time of ♀'s *is always* superior Conjunction (or when she is above *either* the Sun) she is more Easterly than the Sun, *our Morn-* and therefore sets later, and is seen af- *ing or E-* ter Sun-setting, and then she is common- *vening* ly called the *Evening Star*. But from the *Star.* time of her inferior Conjunction till she comes again to the superior, she then appears more Westerly than the Sun, and is only to be seen in the Morning before Sun-rising, and is then called the *Morning Star*.

After the same manner we prove that *Mercury* turns round the Sun, for he always keeps in the Sun's Neighbourhood, and never recedes from him so far as *Venus* does ; and therefore the Orbit of ☿ must lie within that of ♀ : and on the account of his nearness to the Sun, he can seldom be seen without a Telescope.

The Orbit   *Mars* is observed to come in opposition, *of* Mars *in-* and likewise to have all other Aspects with *cludes the* *Earth's.*   the

the Sun; he always preferves a round, full,
and bright Face, except when he is near
his Quadrate Afpect, when he appears fomewhat Gibbous, like the Moon three or four
days before or after the Full: Therefore the
Orbit of ♂ muft include the Earth within
it, and alfo the Sun; for if he was betwixt
the Sun and us, at the time of his inferior
Conjunction, he would either quite difappear, or appear horned, as *Venus* and the *Fig.* 3.
Moon do in that Pofition. Let S be the
Sun, T the *Earth,* and A, P *Mars,* both in
his Conjunction and Oppofition to the Sun,
and in both Pofitions full; and B, C *Mars*
at his Quadratures, when he appears fomewhat gibbous from the Earth at T: Tis
plain hence, that the Orbit of *Mars* does
include the Earth, otherwife he could not
come in oppofition to the Sun; and that it
likewife includes the Sun, elfe he could
not appear full at his Conjunction.

Mars, when he is in oppofition to the
Sun, looks almoft feven times larger in
Diameter, than when he is in conjunction
with him; and therefore muft needs be almoft feven times nearer to us in one pofition
than in the other: For the apparent Magnitudes of far diftant Objects, increafe or de-
creafe in proportion to their diftances from
us. But *Mars* keeps always nearly at the
fame diftance from the Sun; therefore it is
plain, that it is not the Earth, but the Sun
that is the Center of his Motion.   It

12   *The* INTRODUCTION.

It is proved in the fame way, that *Jupiter* and *Saturn* have both the Sun and the Earth within their Orbits; and that the Sun, and not the Earth, is the Center of their Motions; altho' the difproportion of the diftances from the Earth is not fo great in *Jupiter* as it is in *Mars*, nor fo great in *Saturn* as it is in *Jupiter*, by reafon that they are at a much greater diftance from the Sun.

Inferior and Superior Planets   We have now fhewn that all the Planets turn round the Sun, and that *Mercury* and *Venus* are included between him and the Earth; whence they are called the *Inferior Planets*: and that the Earth is placed between the Orbits of *Mars* and *Venus*, and therefore included within the Orbits of *Mars*, *Jupiter* and *Saturn*; whence they are called the *Superior Planets*: And fince the Earth is in the middle of thefe moveable Bodies, and is of the fame nature with them, we may conclude, that fhe has the fame fort of Motions; but that fhe turns round the Sun, is proved thus

The Earth does not ftand ftill, but turns round the Sun.   All the Planets feen from the Earth appear to move very unequally; as fometimes to go fafter, at other times flower; fometimes to go backward, and fometimes to be ftationary, or not to move at all; which could not happen if the Earth ftood ftill.

Fig. 4.   Let S be the Sun, T the Earth, the great Circle A B C D the Orbit of *Mars*, and the

the Numbers 1, 2, 3, &c. its equable Motion
round the Sun ; the correspondent Numbers
1, 2, 3, &c. in the Circle *a, b, c, d*, the Motion of *Mars* as it would be seen from the
Earth.   It is plain from this Figure, that if
the Earth stood still, the Motion of *Mars*
would be always progressive, (tho' sometimes very unequal ;) but since Observations
prove the contrary, it necessarily follows,
that the Earth turns round the Sun.

The Annual Period of the Planets round *The Annu-*
the Sun are determined, by carefully observ- *al and Di-*
urnal Mo-
ing the length of time since their departure *tions of the*
from a certain Point in the Heavens (or from *Planets*
how com-
a Fixed Star) until they arrive to the same *puted.*
again.   By these sorts of Observations the
Antients determined the periodical Revolutions of the Planets round the Sun ; and
were so exact in their Computations, as to
be capable of predicting Eclipses of the Sun
and Moon.   But since the Invention of
Telescopes, Astronomical Observations are
made with greater Accuracy, and of conse-
quence our Tables are far more perfect than
those of the Antients.   And in order to
be as exact as possible, Astronomers compare Observations made at a great distance
of time from one another, including several Periods ; by which means the Error that
might be in the whole, is in each Period
subdivided into such little parts, as to be
very inconsiderable.   Thus the mean Length
of a solar Year is known even to Seconds.

I   The

The Diurnal Rotation of the Planets round their Axes, was difcovered by certain Spots which appear on their Surfaces. Thefe Spots appear firft in the Margin of the Planets Disks, (or the Edge of their Surfaces,) and feem by degrees to creep towards their Middle; and fo on, going ftill forward, till they come to the oppofite Side or Edge of the Disk, where they fet or difappear: and after they have been hid for the fame fpace of time that they were vifible, they again appear to rife, in or near the fame place as they did at firft; then to creep on progreffively, taking the fame courfe as they did before. Thefe Spots have been obferved on the Surfaces of the *Sun, Venus, Mars,* and *Jupiter*; by which means it has been found, that thefe Bodies turn round their own Axes in the times before-mention'd. It is very probable, that *Mercury* and *Saturn* have likewife a Motion round their Axes, that all the parts of their Surfaces may alternately enjoy the Light and Heat of the Sun, and receive fuch changes as are proper and convenient for their Nature. But by reafon of the nearnefs of ☿ to the Sun, and ♄'s immenfe diftance from him, no Obfervations have hitherto been made whereby their Spots (if they have any) could be difcovered, and therefore their diurnal Motions could not be determined. The diurnal Motion of the Earth is computed, from

the

the apparent Revolution of the Heavens, and of all the Stars round it, in the fpace of a natural day. The Solar Spots do not always remain the fame, but fometimes old ones vanifh, and afterwards others fucceed in their room ; fometimes feveral fmall ones gather together, and make one large Spot, and fometimes a large Spot is feen to be divided into many fmall ones. But, notwithftanding thefe Changes, they all turn round with the Sun in the fame time.

The Relative Diftances of the Planets from the Sun, and likewife from each other, are determined by the following Methods: First, the diftances of the two inferior Planets ☿ and ♀ from the Sun in refpect of the Earth's diftance from him, is had by obferving their greateft Elongation from the Sun as they are feen from the Earth. *How the Relative Diftances of the Planets from the Sun are determined.*

The greateft *Elongation* of *Venus* is found by obfervation to be about 48 degrees, which is the Angle S T ♀ ; whence, by the known Rules of Trigonometry, the Proportion of S ♀, the mean Diftance of *Venus* from the Sun to S T, the mean Diftance of the Earth from him, may be eafily found: After the fame manner, in the right-angled Triangle S T ☿, may be found the Diftance S ☿, of *Mercury* from the Sun. And if the mean Diftance of the Earth from the Sun S T be made 1000, the mean Diftance of *Venus* S ♀ from the Sun will be 723 ; *Fig. 5. Elongation.*

and

and of *Mercury* S ☿ 387 : And if the Planets
moved round the Sun in Circles, having
him for their Center, the Diftances here
found would be always their true Diftances;
but as they move in Ellipfes, their Diftances
from the Sun will be fometimes greater, and
fometimes lefs.   Their *Excentricities* are
computed to be as follow, *viz.*

$$\text{Excent. of} \begin{cases} \textit{Mercury } 80 \\ \textit{Venus} \quad \ 5 \\ \textit{Earth} \quad 169 \end{cases} \text{of the parts a-} \\ \text{bovementioned.}$$

The Diftances of the fuperior Planets,
viz. ♂, ♃, and ♄, are found by comparing
their true places, as they are feen from the
Sun, with their apparent places as they are
feen from the Earth.   Let S be the Sun, the
Circle A B C the Earth's Orbit, A G a Line
touching the Earth's Orbit, in which we'll
fuppofe the fuperior Planets are feen from
the Earth, in the Points of their Orbits ♂,
♃, ♄ ; and let D E F G H, be a portion of a
great Circle in the Heavens at an infinite diftance : Then the Place of *Mars* feen from
the Sun is D, which is called his true or
Heliocentrick Place; but from the Earth
he'll be feen in G, which is called his apparent or *Geocentrick Place.*   So likewife
Jupiter and *Saturn* will be feen from the
Sun in the Points E and F, their Heliocentrick Places ; but a Spectator from the
Earth will fee them in the Point of the
Heavens G, which is their Geocentrick Place.

The

Fig.6.

Heliocen-
trick and
Geocentrick
Place,
what.

The Arches D G, E G, F G, the differences between the true and apparent Places of the superior Planets, are called the *Parallaxes* of the Earth's annual Orb, as seen from these Planets. If thro' the Sun we draw S H parallel to A G, the Angles A ♂ S, A ♃ S, A ♄ S, will be respectively equal to the Angles D S H, E S H, and F S H; and the Angle A G S is equal to the Angle G S H, whose measure is the Arch G H; which therefore will be the measure of the Angle A G S, the Angle under which the Semidiameter A S of the Earth's Orbit, is seen from the Starry Heavens. But this Semidiameter is nothing in respect of the immense Distance of the Heavens or Fixed Stars; for from thence it would appear under no sensible Angle, but look like a Point. And therefore in the Heavens the Angle G S H, or the Arch G H vanishes; and the Points G and H coincide; and the Arches D H, E H, F H, may be consider'd as being of the same bigness with the Arches D G, E G, and F G, which are the measures of the Angles A ♂ S, A ♃ S, A ♄ S; which Angles are nearly the greatest Elongation of the Earth from the Sun, if the Earth were observed from the respective Planets, when the Line G ♄ ♃ ♂ A, touches the Earth's Orbit in A. The nearer any of the superior Planets is to the Sun, the greater is the Parallax of the Annual Orb, or the Angle

C under

under which the Semidiameter of the Earth's Orbit is seen from that Planet. In *Mars* the Angle A ♂ S (which is the visible Elongation of the Earth seen from *Mars*, or the Parallax of the Annual Orb seen from that Planet) is about 42 degrees; and therefore the Earth is always to the Inhabitants of *Mars*, either their Morning or Evening Star, and is never seen by them so far distant from the Sun, as we see *Venus*. The greatest Elongation of the Earth seen from *Jupiter*, being nearly equal to the Angle A ♃ S is about 11 degrees. In *Saturn* the Angle A ♄ S is but 6 degrees; which is not much above $\frac{1}{4}$ part of the greatest Elongation we observe in *Mercury*. And since *Mercury* is so rarely seen by us, probably the Astronomers of *Saturn* (except they have better Opticks than we have) have not yet discovered, that there is such a Body as our Earth in the Universe.

The Parallax of the Annual Orb, or the greatest Elongation of the Earth's Orbit seen from any of the superior Planets, being given; the Distance of that Planet from the Sun, in respect of the Earth's Distance from him, may be found by the same Methods as the distances of the inferior Planets were. Thus, to find the Distance of *Mars* from the Sun, it will be as the Sine of the Angle S ♂ A is to the *Radius*, so is the Distance A S, (the Distance of the Earth from

the

the Sun) to S ♂, the Diſtance from the Sun
to *Mars*. After the ſame manner the Diſtances of *Jupiter* and *Saturn* are alſo found.
The mean Diſtance of the Earth from the
Sun being made 1000, the mean Diſtances
of the ſuperior Planets from the Sun are,
viz. the mean Diſtance from the Sun of

$$\begin{Bmatrix} ♂ & 1524 \\ ♃ & 5201 \\ ♄ & 9538 \end{Bmatrix} \text{ and the Excentricity } \begin{Bmatrix} 141 \\ 250 \\ 57 \end{Bmatrix}$$

To which if you add or ſubtract their
mean Diſtances, we ſhall have the greateſt
or leaſt Diſtances of thoſe Planets from the
Sun.

There are other Methods, by which the
relative Diſtances of the Planets might be
found; but that which has been here il-
luſtrated, is ſufficient to evince the Certainty
of that Problem.

Hitherto we have only conſidered the *How the Abſolute Diſtances of the Pla-* Diſtances of the Planets in relation to one
another, without determining them by any
known Meaſure; but in order to find their *nets from the Sun are computed.* abſolute Diſtances in ſome determinate Meaſure, there muſt be ſomething given, whoſe
Meaſure is known. Now the Circumference
of the Earth is divided into 360 Degrees,
and each of theſe Degrees into 60 Geographical Miles, ſo that the whole Circumference contains 21600; and by the known
Proportion for finding the Diameter of a

C 2   Circle

Circle from its Circumference, the Earth's Diameter will be found to be 6872 Miles, and its Semidiameter 3436 Miles. The Parallax of the Earth's Semidiameter, or the Angle under which it is feen from a certain Planet, may be found by comparing the true Place of the Planet as it would be feen from the Center of the Earth, (which is known by computation) with its apparent Place, as it is feen from fome Point on the Earth's Surface. Let C Z A be the Earth, Z C its Semidiameter, ⊕ fome Planet, and B H T an Arch of a great Circle in the Heavens, at an infinite Diftance. Now the Planet ⊕ will appear from the Earth's Center C, in the Point of the Heavens H; but a Spectator from the Point Z, upon the Earth's Surface, will fee the fame Object ⊕ in the Point of the Heavens B; and the Arch B H the Difference is equal to the Angle B ⊕ H = Z ⊕ C, the *Parallax*; which being known, the Side C ⊕ the Diftance of the Planet from the Center of the Earth, at that time, may be eafily found. Now if this Diftance of the Planet from the Earth be determined, when the Centers of the Sun, the faid Planet, and of the Earth, are in the fame right Line, we have the abfolute Diftance of the Planet's Orbit from the Earth's in known meafure: then it will be, as the relative Diftance betwixt the Earth's Orbit and that of the Planet, is to the relative Diftance of the faid Planet from

Parallax of the Earth's Semidiameter.

Fig. 7.

from the Sun; fo is the Diftance of the
Planet's Orbit from the Earth's in known
Meafure, to the diftance of the faid Planet
from the Sun in the fame Meafure: Which
being known, the Diftance of all the other
Planets from the Sun may be found. For,
it will be, as the relative Diftance of any
Planet from the Sun, is to its Diftance from
him in a known Meafure ; fo is the relative
Diftance of any other Planet from him, to
its Diftance in the fame Meafure. This
may be done by finding the Diftance of the
Planet *Mars*, when he is in oppofition to
the Sun, after the fame manner as we find
the Diftance of a Tree, or the like, by two
Stations.

Let ♂ be *Mars*, D the Point on the
Earth's Superficies, where *Mars* is vertical,
when he is in oppofition to the Sun, (which
may be exactly enough found by calculation)
at which time let an Obferver, at the Point
Z, (whofe Situation from D muft be known)
take the Altitude of *Mars*, whofe Complement will be the Angle ♂ Z R ; then in the
Triangle ♂ Z C will be given the Angle
♂ Z C, the Angle C (whofe Meafure is the
Arch D Z) and confequently the Angle
Z ♂ C the Parallax, and alfo the Side Z C
the Semidiameter of the Earth; by which
we may find C ♂ the Diftance of *Mars* from
the Earth. The extreme Nicety required
in this Obfervation, makes it very difficult

C 3   to

to determine the exact Diſtances of the Planets from the Sun; but the celebrated Dr.
Halley has, in the Philoſophical Tranſactions,
ſhewed us a more certain Method for finding the Diſtances of the Planets; which is by
obſerving the Tranſit of *Venus* over the
Sun.

How the Magnitudes of the Planets are determin'd.   The Eye judgeth of the Magnitudes of far
diſtant Objects, according to the Quantities
of the Angles under which they are ſeen
(which are called their apparent Magnitudes;)
and theſe Angles appear greater or leſs in
a certain Proportion to their Diſtances.
Wherefore the Diſtances of the Planets from
the Earth, and their apparent Diameters
being given, their true Diameters, (and from
thence their Magnitudes) may be found.
How the Diſtances of the Planets may be
found, has been already ſhewn; their apparent Diameters are found by a Teleſcope,
having a Machine fixed to it for meaſuring
Fig. 8.   Angles, called a Micrometer. Let B D, or
the Angle B A D be the apparent Diameter
of any Planet, and A B, or A D, (which by
reaſon of the great Diſtances of the Planets
in reſpect of their Magnitudes) may be con-
ſidered as being the Diſtance of the ſaid
Planet from the Obſerver. Now in the
Triangle A B D, having the Sides A B, A D
given, and the Angle A, we have alſo the
other Angles B and D, (becauſe the Sides
A B, A D are equal) whence the Side B D
                                    the

the Diameter of the Planet may be eafily
found by Trigonometry.

From hence it appears, that the fame body
at different Diftances, will feem to have
very different Magnitudes : Thus, the Diameter B D will appear, from the Point E,
to be twice as large as from the Point A.
It alfo follows, that a fmall Body, when
at no great Diftance from us, may appear
to be equal, or even to exceed another at a
great diftance, tho' immenfely bigger. Thus
b d appears under the fame Angle, and confequently of the fame Bignefs from the
Point A, that the Line B D doth, tho' one
vaftly exceeds the other. And this is the *Why the*
Reafon, why the Moon, which is much lefs *Moon ap-*
pears big-
than any of the Planets, appears to us vaftly *ger than*
bigger than either of them, and even to equal *any of the*
Planets.
the Sun himfelf, which is many thoufand
times greater in Magnitude.

The Diftances of the Planets, and Periods
round the Sun, their Diameters and Velo-
cities round their own Axes, according to
modern Computations, are as follow.

Saturn

		Y.	D.	H.	Distance in Miles.
Saturn		29	167	22	777.000.000
Jupiter		11	314	12	424.000.000
Mars	Revolves about the Sun in the Space of	1	321	23	123.000.000
Earth		0	365	6	81.000.000
Venus		0	224	16	59.000.000
Mercury		0	87	23	32.000.000

		D.	H.	M.	
Moon	Round the Earth	27	7	43	240 : 000

	Periods round their own Axes.			Diameters in Miles
	D.	H.	M.	
Sun	25	6	0	763.000
Saturn	.	. . .	. .	61.000
Jupiter	0	9	56	81.000
Mars	1	0	40	4.440
Earth	0	23	56	7.970
Venus	0	23	0	7.900
Mercury	.	. . .	. .	4.240
Moon	27	7	43	2.170

The Cause of *Eclipses*, and *Phases* of the Moon, and some other Phenomena not here explained, shall be shewed when we come to give a Description of the *Orrery*.

Besides the Planets already mentioned, there are other great Bodies that sometimes visit our System, which are a sort of Temporary Planets; for they come and abide
with

Plate 2.

Page 24

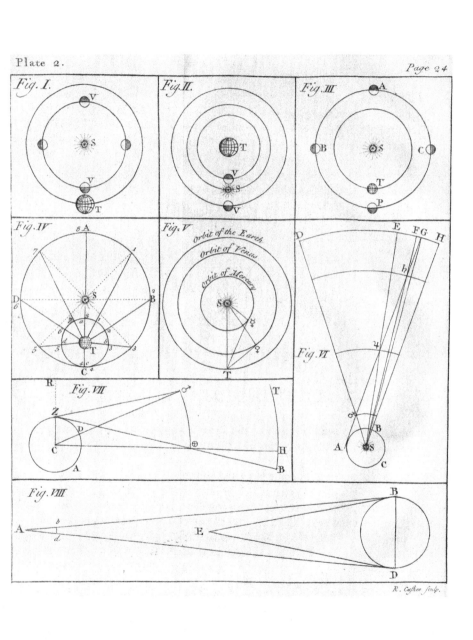

Fig. I.

Fig. II.

Fig. III.

Fig. IV.

Fig. V.

Orbit of the Earth

Orbit of Venus

Orbit of Mercury

Fig. VI.

Fig. VII

Fig. VIII

R. Cushee sculp.

with us for a while, and afterwards with-
draw from us, for a certain fpace of Time,
after which they again return. Thefe wandring Bodies are called *Comets*.

The Motion of Comets in the Heavens, *of Comets.*
according to the beft Obfervations hitherto
made, feem to be regulated by the fame
immutable Law that rules the Planets; for
their Orbits are Elliptical, like thofe of the
Planets, but vaftly narrower, or more Excentrick. Yet they have not all the fame
Direction with the Planets, who move from
Weft to Eaft, for fome of the Comets move
from Eaft to Weft; and their Orbits have
different Inclinations to the Earth's Orbit;
fome inclining Northwardly, others South-
wardly, much more than any of the Planetary Orbits do.

Altho' both the Comets and the Planets
move in Elliptick Orbits, yet their Motions
feem to be vaftly different; for the Excentricities of the Planet's Orbits are fo fmall,
that they differ but little from Circles;
but the Excentricities of the Comets are
fo very great, that the Motions of fome of
them feem to be almoft in right Lines, tend-
ing directly towards the Sun.

Now, fince the Orbits of the Comets are
fo extremely Excentrick, their Motions,
when they are in their *Perihelion*, or neareft
Diftance from the Sun, muft be much fwifter
than when they are in their *Aphelion*, or
fartheft Diftance from him; which is the
reafon

reaſon why the Comets make ſo ſhort a
ſtay in our Syſtem; and when they diſap-
pear, are ſo long in returning.

The Figures of the Comets are obſerved
to be very different; ſome of them ſend
forth ſmall Beams like Hair every way round
them; others are ſeen with a long fiery
Tail, which is always oppoſite to the Sun.
Their Magnitudes are alſo very different,
but in what Proportion they exceed each
another, is as yet uncertain. It is not probable, that their Numbers are yet known,
for they have not been obſerved, with due
Care, nor their Theories diſcovered, but of
late Years. The Antients were divided in
their Opinions concerning them; ſome imagined that they were only a kind of *Me-
teors* kindled in our Atmoſphere, and were
there again diſſipated; others took them to
be ſome ominous Prodigies. But modern
Diſcoveries prove, that they are Worlds
ſubject to the ſame Laws of Motion as the
Planets are: and they muſt be very hard
and durable Bodies, elſe they could not bear
the vaſt Heat, which ſome of them, when
they are in their *Perihelia,* receive from the
Sun, without being utterly conſumed. The
great Comet which appear'd in the Year 1680,
was within $\frac{1}{6}$ part of the Sun's Diameter from
his Surface; and therefore its Heat muſt be
prodigiouſly intenſe beyond imagination.
And when it is at its greateſt diſtance from
the Sun, the Cold muſt be as rigid.

SECT

SECT. II.

Of the FIXED STARS.

THE fixed Stars are thofe bright and fhining Bodies, which in a clear Night appear to us every where difperfed through the boundlefs Regions of Space. They are term'd fixed, becaufe they are found to keep the fame immutable Diſtance one from another in all Ages, without having any of the Motions obferved in the Planets. The *The fixed* fixed Stars are all placed at fuch immenfe *Stars are* Diftances from us, that the beft of Tele- *at immenfe* *diſtance* fcopes reprefent them no bigger than Points, *from us.* without having any apparent Diameters.

It is evident from hence, that all the Stars *The fixed* are luminous Bodies, and fhine with their *Stars are* *luminous* own proper and native Light, elfe they *Bodies, like* could not be feen at fuch a great diftance. *the Sun.* For the *Satellits* of *Jupiter* and *Saturn*, tho' they appear under confiderable Angles through good Telefcopes, yet are altogether invifible to the naked Eye. *The Dif-* Although the Diftance betwixt us and the *tance from* *us to the* Sun is vaftly large when compared to the *Sun is no-* Diameter of the Earth, yet it is nothing *thing in* *comparifon* when compared with the prodigious Diftance *of the vaſt* of the fixed Stars; for the whole Diameter *Diſtance of* *the fixed* *of Stars.*

of the Earth's Annual Orbit, appears from the neareſt fixed Star no bigger than a Point, and the fixed Stars are at leaſt 100,000 times farther from us than we are from the Sun; as may be demonſtrated from the Obſervations of thoſe who have endeavoured to find the Parallax of the Earth's Annual Orb, or the Angle under which the Earth's Orbit appears from the fixed Stars.

As to ap-pearance, the Earth maybe con-ſidered as being in the Centre of the Hea-vens.

Hence it follows, that tho' we approach nearer to ſome fixed Stars at one time of the Year than we do at the oppoſite, and that by the whole Length of the Diameter of the Earth's Orbit ; yet this Diſtance being ſo ſmall in compariſon with the Diſtance of the fixed Stars, their Magnitudes or Poſitions cannot thereby be ſenſibly altered. Therefore we may always, without Error, ſuppoſe ourſelves to be in the ſame Centre of the Heavens, ſince we always have the ſame viſible Proſpect of the Stars without any Alteration.

The fixed Stars are Suns.

If a Spectator was placed as near to any fixed Star, as we are to the Sun, he would there obſerve a Body as big, and every way like, as the Sun appears to us ; and our Sun would appear to him no bigger than a fixed Star: and undoubtedly he would reckon the Sun as one of them in numbring the Stars. Wherefore ſince the Sun differeth nothing from a fixed Star, the fixed Stars may be reckoned ſo many Suns.

It

It is not reasonable to suppose that all the fixed Stars are placed at the same distance from us; but it is more probable that they *The fixed* are every were· interspersed thro' the vast *Stars are* indefinite Space of the Universe; and that *Distance* there may be as great a Distance betwixt any *from each* two of them, as there is betwixt our Sun and *other.* the nearest fixed Star. Hence it follows, why they appear to us of different Magnitudes, not because they really are so, but because they are at different Distances from us; those that are nearest, excelling in Brightness and Lustre those that are more remote, who give a fainter Light, and appear smaller to the Eye.

The Astronomers distribute the Stars into se-veral Orders or Classes; those that are nearest *The Distri-* to us, and appear brightest to the Eye, are *bution of* called Stars of the first Magnitude; those *the Stars* that are nearest to them in Brightness and *Classes:* Lustre, are called Stars of the second Magni-tude; those of the third Class, are stiled Stars of the third Magnitude; and so on, until we come to the Stars of the sixth Magnitude, which are the smallest that can be discerned by the naked Eye. There are infinite num-bers of smaller Stars, that can be seen through Telescopes; but these are not reduced to any of the six Orders, and are only called *Telescopical* Stars. It may be here observed, *Of Telesco-* that tho the Astronomers have reduced all *pical Stars.* the Stars that are visible to the naked Eye, into some one or other of these Classes; yet

3   we

we are not to conclude from thence that all the Stars answer exactly to some or other of these Orders; but there may be in reality as many Orders of the Stars as they are in Number, few of them appearing exactly of the same Bigness and Lustre.

The antient Astronomers, that they might distinguish the Stars, in regard to their Situation and Position to each other, divided the whole starry Firmament into several *Asterisms*, or Systems of Stars, consisting of those that are near to one another. *The Stars digested into Constellations.* These *Asterisms* are called *Constellations*, and are digested into the Forms of some Animals, as Men, Lions, Bears, Serpents, &c. or to the Images of some known things, as of a Crown, a Harp, a Triangle, &c.

The Starry Firmament was divided by the Antients into 48 Images or Constellations; twelve of which they placed in that part of the Heavens wherein are the Planes of the Planetary Orbits; which part is *Zodiack.* called the *Zodiack*, because most of the Constellations placed therein resemble some living Creature. The two Regions of the Heavens that are on each side of the *Zodiack*, are called the North and South Parts of the Heavens.

Constellations within the Zodiack. The Constellations within the *Zodiack*, are, 1. *Aries*, the *Ram*; 2. *Taurus*, the *Bull*; 3. *Gemini*, the *Twins*; 4. *Cancer*, the *Crab*; 5. *Leo*, the *Lion*; 6. *Virgo*, the *Virgin*;

Virgin ; 7. *Libra*, the *Ballance* ; 8. *Scorpio*, the *Scorpion* ; 9. *Sagittarius*, the *Archer* ; 10. *Capricornus*, the *Goat* ; 11. *Aquarius*, the *Water-Bearer* ; and, 12. *Pisces*, the *Fishes*.

The Conftellations on the North Side *Northern Conftellations.* of the *Zodiack* are Twenty One, *viz.* the *Little Bear* ; the *Great Bear* ; the *Dragon* ; *Cepheus*, a King of *Ethiopia* ; *Boötes*, the Keeper of the *Bear* ; the *Northern Crown* ; *Hercules* with his Club, watching the *Dragon* ; the *Harp* ; the *Swan* ; *Caffiopeia* ; *Perfeus* ; *Andromeda* ; the *Triangle* ; *Auriga* ; *Pegafus*, or the *Flying Horfe* ; *Equuleus* ; the *Dolphin* ; the *Arrow* ; the *Eagle* ; *Serpentarius* ; and the *Serpent*.

The Conftellations noted by the Antients *Southern Conftellations* on the South Side of the *Zodiack*, were fifteen, *viz*. The *Whale*, the River *Eridanus*, the *Hare* ; *Orion* ; the *Great Dog* ; *Little Dog* ; the Ship *Argo* ; *Hydra* ; the *Centaur* ; the *Cup* ; the *Crow* ; the *Wolf* ; the *Altar* ; the *Southern Crown* ; and the *Southern Fifh*. To thefe have been lately added the following, *viz*. The *Phenix* ; the *Crane* ; the *Peacock* ; the *Indian* ; the *Bird of Paradife* ; the *Southern Triangle* ; the *Fly* ; *Cameleon* ; the *Flying Fifh* ; *Toucan*, or the *American Goofe* ; the *Water Serpent*, and the *Sword Fifh*. The Antients placed thefe particular Conftellations or Figures in the

3   Heavens,

Heavens, either to commemorate the Deeds of some great Man, or of some notable Exploit or Action; or else took them from the Fables of their Religion, &c. And the Modern Astronomers do still retain them, to avoid the Confusion that would arise by making new ones, when they compared the modern Observations with the old ones.

Some of the principal Stars have particular Names given them, as *Syrius*, *Arcturus*, &c. There are also several Stars that are not reduced into Constellations, and *Unformed Stars.* these are called *Unformed Stars*.

Besides the Stars visible to the naked Eye, there is a very remarkable Space in the Heavens, called the *Galaxy*, or *Milky Way*. *The Galaxy or Milky Way.* This is a broad Circle of a whitish Hue, like Milk, going quite round the whole Heavens; and consisting of an infinite Number of small Stars, visible thro' a Telescope, tho' not discernible by the naked Eye, by reason of their exceeding faintness; yet with their Light they combine to illustrate that part of the Heavens where they are, and to cause that shining Whiteness.

The Places of the Fixed Stars, or their relative Situations one from another, have been carefully observed by Astronomers, and digested into Catalogues. The first among the *Greeks*, who reduced the Stars into a Catalogue, was *Hypparcus*, who, from his own Observations, and of those who lived

lived before him, inferted 1022 Stars into
his Catalogue, about 120 Years before the
Chriftian *Æra:* This Catalogue has been
fince enlarged and improved, by feveral
learned Men, to the Number of 3000; of
which there are a great many Telefcopical,
and not to be difcerned by the naked Eye;
and thefe are all ranked in the Catalogue,
as Stars of the feventh Magnitude.

It may feem ftrange to fome, that there are
no more than this Number of Stars vifible
to the naked Eye; for fometimes in a clear
Night, they feem to be innumerable. But
this is only a Deception of our Sight, arifing
from their vehement fparkling, while we
look upon them confufedly, without re-
ducing them into any Order; for there can
feldom be feen above 1000 Stars in the
whole Heavens, with the naked Eye at the
fame time; and if we fhould diftinctly view
them, we fhall not find one but what is
inferted, upon a good *Celeftial* Globe.

Altho' the Number of Stars that can be
difcerned by the naked Eye, are fo few, yet
it is probable there are many more which
are beyond the reach of our Opticks; for
thro' Telefcopes they appear in vaft Multi-
tudes, every where difperfed throughout the
whole Heavens; and the better our Glaffes
are, the more of them we ftill difcover. The
ingenious Dr. *Hook* has obferved 78 Stars in
the *Pleiades,* of which the naked Eye is

never

never able to difcern above 7; and in *O-rion*, which has but 80 Stars in the *Britifh* Catalogue, (and fome of them Telefcopical) there has been numbered 2000 Stars.

An Idea of the Univerfe.   Thofe who think that all thefe glorious Bodies were created for no other purpofe, than to give us a little dim Light, muft entertain a very flender Idea of the Divine Wifdom; for we receive more Light from the *Moon* it felf, than from all the *Stars* put together. And fince the *Planets* are fubject to the fame Laws of Motion with our *Earth*, and fome of them not only equal, but vaftly exceed it in Magnitude, it is not unreafonable to fuppofe, that they are all habitable Worlds. And fince the *Fixed Stars* are no ways behind our *Sun*, either in Bignefs or Luftre; is it not probable, that each of them have a Syftem of *Planetary Worlds* turning round them, as we do round our Sun. And if we afcend as far as the fmalleft Star we can fee, fhall we not then difcover innumerable more of thefe glorious Bodies, which now are altogether invifible to us? and fo *ad infinitum*, thro' the boundlefs Space of the Univerfe. What a magnificent Idea muft this raife in us of the *Divine Being!* Who is every where, and at all Times prefent, difplaying his Divine Power, Wifdom and Goodnefs amongft all his Creatures!

The

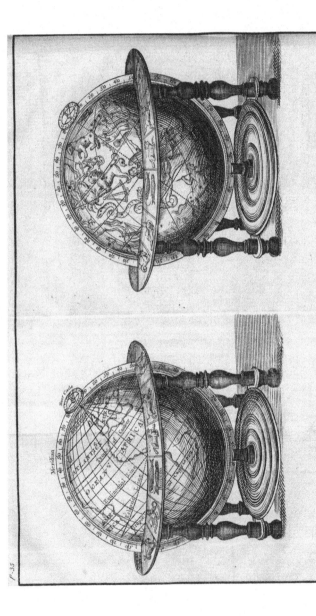

NEW and CORRECT GLOBES according to the latest Observations

Made and Sold by RICHARD CUSHEE at the Globe and Sun between St Dunstans Church & Chancery Lane

FLEET STREET

ESTATES Surveyd *also* MAPS DRAWN and ENGRAV'D

R. Cushee sculp.

P. 35

The DESCRIPTION and USE of the CELESTIAL and TERRESTRIAL GLOBES.

Globe or *Sphere* is a round folid *Sphere or Globe.* Body, having every part of its Surface equally diftant from a Point within it, called its *Center*; and it may be conceived to be formed by the Revolution of a Semicircle round its Diameter.

Any Circle paffing through the Center *Great Circle.* of the Sphere, thereby dividing into two equal Parts or Segments, is called a *Great Circle*; and the Segments of the Sphere fo divided, are called *Hemifpheres.* *Hemifpheres.*

Every Great Circle has its Poles and Axis.

The *Poles* of a Great Circle, are two *Poles.* Points on the Surface of the Sphere diametrically oppofite to one another, and every where equally diftant from the faid Circle.

The *Axis* of a Circle is a right Line *Axis.* paffing through the Center of the Sphere, and through the Poles of the faid Circle; and is therefore perpendicular to the Plane thereof.

All Circles paffing through the Poles of any great Circle, interfect it in two Places diametrically oppofite, and alfo at right Angles; and with refpect to the faid Great Circle, they may be called its *Secundaries*.

All Circles dividing the Sphere into two unequal Parts, are called *leffer* or *parallel Circles*, and are ufually denominated by that Great Circle, to which they are parallel.

The Earth being globular, its outward Parts, as the feveral *Countries*, *Seas*, *&c.* are beft, and moft naturally reprefented upon the Superficies of a Globe; and when fuch a Body has the outward Parts of the Earth and Sea delineated upon its Surface, and placed in their natural Order and
Situation, it is called a *Terreftrial Globe*.

The Celeftial Bodies appear to us as if they were all placed in the fame Concave Sphere; therefore Aftronomers place the Stars according to their refpective Situations and Magnitudes; and alfo the Images of the Conftellations, upon the external Surface of a Globe; for it anfwers the fame purpofes as if they were placed within a concave Sphere, if we fuppofe the Globe to be tranfparent, and the Eye placed in the Center. A Globe having the Stars placed upon its Surface, as above defcribed, is
called a *Celeftial Globe*. Thefe Globes are both placed in Frames, with other Appurtenances, as fhall be defcribed in a proper place.   The

The principal Ufes of thefe Globes, (befides their ferving as *Maps* to diftinguifh
the outward Parts of the Earth, and the Situations of the Fixed Stars) is to explain and
refolve the Phænomena arifing from the
diurnal Motion of the Earth round its Axis.

It has been fhewed in the Introduction,
that the Diftance of the Earth from the Sun,
is no more than a Point, when compared
with the immenfe Diftance of the Fixed
Stars; therefore let the Earth be in what
Point foever of her Orbit, there will be the
fame Profpect of the Heavens, as a Spectator
would obferve did he refide in the Sun :
and if feveral Circles be imagined to pafs
thro' the Center of the Earth, and others,
parallel to them, be conceived to pafs thro'
the Center of the Sun, thefe Circles in the
Heavens will feem to coincide, and to pafs
exactly thro' the fame Stars. Wherefore as
to the Appearances of the Fixed Stars, it is
indifferent whether the Earth or the Sun be
made the Center of the Univerfe. But
becaufe it is from the Earth that we always
obferve the Celeftial Bodies, and their apparent Motions feem to us to be really
made in the Heavens, it is more natural in
explaining the Phænomena arifing from
thefe Motions, to place the Earth in the
Center. And again, becaufe the Semidiameter of the Earth, when compared to her
Diftance from the Sun, is of no fenfible

margin notes: *The princi-pal Ufe of the Globes. There will be the fame Profpect of the Fixed Stars, whe-ther the Spectator be placed on the Earth or in the Sun.*

D 3   Mag-

Magnitude, any Point upon the Earth's Surface, let her be in what Part foever of her Orbit, may be confidered as being the Center of the Univerfe. Upon thefe Principles, the different Phænomena arifing from the diurnal Motion of the Earth, and the different Situation of a Spectator upon its Surface, are very naturally illuftrated and explained by the Globes.

As to the Alterations of Seafons, *&c.* arifing from the annual Motion of the Earth round the Sun, it is indifferent which we fuppofe to move, the Earth, or the Sun; for in both Cafes the Effect will be the fame: Wherefore becaufe it is the Sun, that appears to us to move, we fay the Sun is in fuch a Part of the Ecliptick, without attributing any Motion to the Earth, any more than if fhe had actually been at reft. For the fame reafon we fay, the Sun rifes, or the Sun fets; by which we mean, that he begins to appear or difappear, without confidering in the leaft how thefe Effects are produced. Thefe things are here mentioned, to obviate the Objections that are fometimes made by Beginners, after they had been told that the Sun ftands ftill.

SECT.

✠✠✠✠✠✠✠✠✠✠✠✠✠

SECT. I.

An Explanation of the Circles of the Sphere, and of some Astronomical Terms arising therefrom.

IN order to determine the relative Situations of Places upon the Earth, as well as the Positions of the Fixed Stars, and other Celestial Phænomena; the Globe of the Earth is supposed to be environ'd by several imaginary Circles, and these are called, the *Circles of the Sphere.* These *The Circles of the Sphere.* imaginary Circles are either fixed, and always obtain the same Position in the Heavens; or moveable, according to the Position of the Observer.

Those Circles that are fixed, owe their Origin to the twofold Motion of the Earth; and are the *Equator,* the *Ecliptick,* with their *Secundaries,* and *Parallels.* These fixed Circles are usually delineated upon the Surface of the Globes.

The moveable Circles are only the *Horizon,* its *Secundaries* and *Parallels:* These are represented by the Wooden Frame, and the Brass Ring, wherein the Globe is hung, and a thin Plate of Brass to be screwed in a proper Place upon the said Ring, as Occasion requires.

I. Of the Equinoctial.

The Equator or E-quinoctial. I. The *Equator* or the *Equinoctial*, is that Great Circle in the Heavens, in whose Plane the Earth performs its diurnal Motion round its Axis ; or it is that Great Circle, parallel to which the whole Heavens seem to turn round the Earth from East to West in 24 Hours.

Note, The Equator and the Equinoctial are generally synonimous Terms ; but sometimes the Equator particularly signifies that Great Circle upon the Surface of the Earth, which coincides with the Equinoctial in the Heavens. This Circle is also by Mariners commonly called the *Line*.

Northern and Southern Hemispheres. The Equinoctial divides the Globe of the Earth, and also the whole Heavens into two equal Parts, North and South, which are called the *Northern* and *Southern Hemispheres*. The Axis of this Circle is called *The Axis of the World.* the *Axis of the World*, or the *Earth's Axis*, because the Earth revolves about it (from West to East) in 24 Hours. The Extremes *Poles of the World, or of the E-quator.* of this Axis are called the *Poles* of the *World*, whereof that which lies in the Northern Hemisphere, is called the *North Pole*, and the other is called the *South Pole*.

The

The Equinoctial Circle is always deline-ated upon the Surface of each Globe, with its Name at length expreſſed ; the Axis of this Circle or the Earth's Axis, is only an imaginary Line in the Heavens, but on the Globes it is expreſſed by the Wires about which they really turn. The Poles of the World are the two Points upon the Surface of the Globe through which theſe Wires paſs ; the North Pole is that which hath the little braſs Circle, with a moveable Index placed round it ; and the other oppoſite to it, is the South Pole. The Northern Hemi-ſphere is that wherein the North Pole is placed, and the oppoſite one is the Southern Hemiſphere.

The Aſtronomers divide all Circles into 360 equal Parts called *Degrees,* each De-gree into 60 equal Parts called *Minutes,* each Minute into 60 Seconds, *&c.* But be-ſides this Diviſion into Degrees, the Equi-noctial is commonly divided into 24 equal Parts or *Hours,* each Hour into 60 *Minutes,* each Minute into 60 *Seconds, &c.* ſo that one Hour is equal to 15 Degrees, each Mi-nute of Time is equal to 15 Minutes of a De-gree, *&c.*

2. All Circles conceived to paſs through the Poles of the World, interſecting the Equinoctial at Right Angles, are with re-ſpect to any Point in the Heavens call'd *Hour Circles*; and alſo *Circles* of *Aſcenſion,* be-

Hour Cir-cles, or Cir-cles of A-ſcenſion, al-ſo called Meridians.

becaufe the Afcenfion of the Heavenly Bodies, from a certain Point, are by them determined.

Thefe Circles are alfo with regard to
Places upon the Earth, call'd *Meridians*.

The *Meridians* are commonly drawn upon the Terreftrial Globe thro' every 15 Degrees of the Equinoctial, thereby making an
Hour difference betwixt the places through
which they pafs. On the Celeftial Globe
there are commonly drawn but two of
thefe *Meridians* croffing the Equinoctial in
four Points equidiftant from one another,
thereby dividing it into four Quadrants ;
but the intermediate ones are here fupplied,
and alfo upon the Terreftrial Globe, by the
Brafs Circle in which they are hung, which
The Brafs is therefore called the *Brafs Meridian*, and
Meridian. fometimes only the *Meridian* ;. it ferving for
this purpofe to all the Points upon either
Globe.

There is alfo a little Brafs Circle fixed upon this Meridian, divided into 24 Hours,
having an Index moveable round the Axis of
the Globe to be turned to any particular
Hour. The Ufe of this Circle is to fhew
the difference of Time betwixt any two
The Hour Meridians, and is therefore called the *Hour*
Circle. *Circle.*

3. All Circles parallel to the Equinoctial,
are with refpect to any Point in the Hea-
Parallels vens, called *Parallels* of *Declination*. So
of Declina- that,   4. The
tion.

4. The *Declination* of any *Point* in the Heavens (as of the *Sun*, a *fixed Star*, or the like) is an Arch of the Meridian paſſing through that Point, and intercepted betwixt it and the Equator : and if the ſaid Point be to the $\left\{\begin{array}{c}\text{Northward}\\\text{Southward}\end{array}\right\}$ of the Equator, it is called $\left\{\begin{array}{c}North\\South\end{array}\right\}$ *Declination.*

Declination North or South.

Of the Parallels of Declination, four are eminently diſtinguiſhed by particular Names, *viz.* The two *Tropicks*, and the two *Polar Circles.*

Tropicks and Polar Circles.

5. The Tropicks are on different ſides of the Equator, each 23 Degrees and 29 Minutes diſtant from it ; that which lies in the Northern Hemiſphere, is called the *Tropick of Cancer* ; and the Southern one, the *Tropick of Capricorn.*

Tropick of Cancer ; of Capricorn.

Theſe Circles are the Limits of the Sun's greateſt Declination, and are called Tropicks, becauſe whenever the Sun arrives to them, he ſeems to return back again towards the Equator.

6. The *Polar Circles* are each of them at the ſame diſtance from the Poles of the World, that the Tropicks are from the Equator, *viz.* 23° 29'. That which lies near the North Pole, is called the *Arctick Circle,* from *Arctos* a Conſtellation ſituated in the Heavens near that Place; whence alſo this Pole is ſometimes called the *Arctick Pole.*

Arctick Circle.

Arctick Pole.

The

The other Polar Circle which is fituated near the South Pole, is called the *Antarctick Circle*; becaufe its Pofition is contrary to the other : And the South Pole is fome- *Antarctick* times called the *Antarctick Pole*.
Circle.   The Tropicks and the Polar Circles have each their Names expreffed upon the Globes.

II. *Of the Ecliptick.*

Ecliptick.   7. The *Ecliptick* is that great Circle in whofe Plane the Earth performs its annual Motion round the Sun ; or, in which the Sun feems to move round the Earth once in a Year : This Circle makes an Angle with the Equinoctial of 23 Degrees 29 Minutes ; and interfects it in two oppofite Points, which *Equinoc-* are called the *Equinoctial Points* ; and the *tial Points.* two Points in the Ecliptick that are at the greateft diftance from the Equinoctial Points, *Solftitial* are called the *Solftitial Points.* The two *Points.* Meridians paffing through thofe Points, *Colures.* are, by way of Eminence, called *Colures* ; whereof that which paffeth thro' the Equi- *Equinoc-* noctial Points is called the *Equinoctial Co-* *tialColure.* *lure* ; and that which is at Right Angles to it, paffing through the Solftitial Points, is *Solftitial* called the *Solftitial Colure.* *Colure.*
The Eclip-   The Ecliptick is divided into 12 equal *tick divi-* Parts called *Signs*, each Sign being 30 De- *ded into* *Signs.* grees, beginning from one of the Equinoctial Points, and numbered from Weft to Eaft:

the Names and Characters of the 12 Signs are as follow : *viz.*

Aries, Taurus, Gemini, Cancer, Leo. Virgo;
1. ♈   2. ♉   3. ♊   4. ♋ 5. ♌ 6. ♍
Libra, Scorpio, Sagittarius, Capricornus, Aquarius, Pisces.
7. ♎   8. ♏   9. ♐   10. ♑   11. ♒   12. ♓

The firſt ſix of theſe are called the *Nor-* _{Northern} *thern Signs*, and poſſeſs that half of the _{Signs.} Ecliptick which is to the Northward of the Equator ; beginning with the firſt Point of ♈, and ending with the laſt Point of ♍.

The latter Six are called the *Southern* _{Southern} *Signs*, becauſe they poſſeſs the Southern _{Signs.} half of the Ecliptick, beginning at the firſt Point of ♎, and ending with the laſt Point of ♓.

The Diviſion of the Ecliptick into Signs, and the Names of the Colures, are particularly expreſſed upon the Globes.

The Signs of the Ecliptick took their Names from 12 Conſtellations mentioned in the Introduction to be ſituated in the Heavens near thoſe Places. It is to be obſerved, that the Signs are not to be confounded with the Conſtellations of the ſame Name : For the *Sign* of *Aries* is not the ſame with the *Conſtellation* Aries ; the latter is a Syſtem of Stars digeſted into the Figure of a *Ram* ; but the Sign of Aries is only 30 Degrees of the Ecliptick counted from the Equinoctial Point ♈, (which is reckon'd the firſt Point in the Ecliptick) to the beginning of *Taurus* : Or, it is ſometimes taken for all that ſpace upon the Celeſtial

leftial Globe contained between the two
Circles paffing through the firft Points of
♈ and ♉. What has been here faid of *Aries*,
is to be noted of all the reft of the Signs.

The Conftellations above mentioned were
formerly fituated within the Signs which
now bear their Names; but by a flow Motion of the Equinoctial Points, being one
Degree in 72 Years, the Conftellation *Aries*
has now got into the Sign ♉, and fo of the
reft. So that *Pifces* is now got into the
Sign of ♈; this flow Motion in the Heavens
is called the *Preceffion of the Equinoxes.*
By this Motion, that Star which we now
call the *Pole Star* will in procefs of time be
to the Southward of *London.*

Poles of the
Ecliptick. The *Poles of the Ecliptick* are both fituated in the Solftitial Colure, at 23 Degrees,
29 Minutes diftance from the Pole of the
World; and they take their Denomination
from the Hemifphere wherein they are placed,
viz. that which lies in the $\left\{ \begin{array}{c} \text{Northern} \\ \text{Southern} \end{array} \right\}$ He-

mifphere, is called the $\left\{ \begin{array}{c} \text{North} \\ \text{South} \end{array} \right\}$ Pole of

the Ecliptick. The Arctick and Antarctick
Circles are defcribed by the Poles of the
Ecliptick in the Diurnal Motion of the
Earth round its Axis, whence it feems thefe
two Circles are called *Polar.*

8. All Great Circles paffing through the
Poles of the Ecliptick, and confequently

in-

interfecting it at right Angles) are called *Circles of Longitude.*
Circles of *Longitude*: So that

9. The *Longitude* of any *Point* in the *Longitude of any Point in the Heavens.* Heavens, (as a *Star* or *Planet*, &c.) is an Arch of the Ecliptick contained between the Circle of Longitude paffing thro' that Point, and the Equinoctial Point ♈. And that degree of any Sign which lies under the Circle of Longitude, paffing thro' any Star or Planet, is called the *Place* of that Star or *Place of a Star.* Planet.

Note, The *Sun* never goes out of the Ecliptick, and it is not ufual to fay the Sun's Longitude, but we commonly exprefs it the *Sun's Place*, which is that Sign, Degree, Minute, &c. of the Ecliptick, which he at any time poffeffes.

10. All Circles conceived to be drawn parallel to the Ecliptick, are called *Parallels* of *Latitude*: So that,

11. The *Latitude* of any Point in the *Latitude of a Star, &c.* Heavens, (as a Fixed Star, &c.) is an Arch of the Circle of Longitude, paffing thro' that Point, and intercepted betwixt it and the Ecliptick; or, the Latitude is the Diftance from the Ecliptick: And if the faid Point be to the Northward of the Ecliptick, it is called North Latitude; but if it be to the Southward, it is called South Latitude.

Upon the *Terreftrial Globe* none of the Circles of Longitude are defcribed; and upon the *Celeftial*, they are commonly
drawn

drawn thro' the beginning of every *Sign*; but they are all ſupplied upon both Globes, by faſtening a thin Plate of Braſs over one of the Poles of the Ecliptick, and ſo as to be moved to any Degree thereof at pleaſure. The Parallels of Latitude are alſo ſupplied by the Graduations upon the ſaid Plate, as ſhall be ſhewed in a proper place.

We have now done with all thoſe Circles that are Fixed, and ſuch as are drawn upon the Globes themſelves; we next proceed to the Moveable Circles.

III. *Of the Horizon.*

Horizon.　12. The *Horizon* is that Great Circle which divides the upper or viſible Hemiſphere of the World, from the lower or inviſible: This Circle is diſtinguiſhed into two Sorts, the *Senſible* and the *Rational.*

Senſible Horizon.　The *Senſible* or *Apparent Horizon* is that Circle which limits or determinates our Proſpect, whether we are at Land or Sea, reaching as far as we can ſee; or it is that Circle where the Sky, and the Earth or Water ſeem to meet. When we are on *Terra Firma*, this Circle commonly ſeems rugged and irregular, occaſion'd by the unevenneſs of the Ground terminating our Proſpect; but at Sea there are no ſuch Irregularities. The Semidiameter of this Circle varieth according to the heighth of the Eye

of

of the Obferver; if a Man of fix foot high ftood upon a large Plain, or the Surface of the Sea, he could not fee above three Miles round.

This Circle determines the Rifing and Setting of the Heavenly Bodies, and diftinguifhes Day and Night.

The *Rational* or True *Horizon* is a great *Rational Horizon.* Circle paffing thro' the Center of the Earth parallel to the fenfible Horizon, being diftant from it by the Earth's Semidiameter, which is about 3980 Miles: This Diftance is nothing in comparifon of the immenfe Diftance of the Sun and the Fixed Stars, therefore Aftronomers make no diftinction between thefe two Circles, but confider the apparent Horizon, or that wherein the Sun appears to rife and fet, as paffing thro' the Center of the Earth.

This Circle is divided by Aftronomers into four Quadrants, and each of thefe Quadrants into 90 degrees, &c. The four Points quartering this Circle are called the *Cardinal Points,* and are termed the *Eaft, Weft, North* and *South.* The *Eaft* is that Point *Cardinal Points of the Horizon.* of the Horizon where the Sun rifes when he is in the Equinoctial, or on that Day when he afcends above the Horizon exactly at fix o' Clock; and the *Weft* is that Point of the Horizon which is directly oppofite to the Eaft, or where the Sun fets when he is in the Equinoctial. The *South* is 90 degrees diftant from the Eaft and Weft, and

E   is

is towards that part of the Heavens wherein
the Sun always appears to us in *Great Britain* at Noon ; and the *North* is that part
of the Heavens which is directly opposite
to the South. Or the North and South
Points of the Heavens may be found, by
turning your self directly either towards the
East or the West : If you look towards the
$\begin{Bmatrix} \text{East} \\ \text{West} \end{Bmatrix}$ the $\begin{Bmatrix} \text{South} \\ \text{North} \end{Bmatrix}$ will be to the Right
Hand, and the $\begin{Bmatrix} \text{North} \\ \text{South} \end{Bmatrix}$ to the Left.

*Points of
the Com-
pass.*

Besides the aforementioned Division of
the Horizon into Degrees, *Mariners* divide
it into 32 equal Parts, which they call the
Points of the *Compass* ; to each of which
Points they give a particular Name, com-
pounded of the four Cardinals, according to
what Quarter of the Compass is intended.

The Center of the Horizon is the Place
of Observation, and the Poles of it are, one

Zenith.

exactly over our Heads, called the *Zenith* ;
and the other exactly under our Feet, called

Nadir.

the *Nadir.*

*Vertical
Circles.*

13. All Circles conceived to pass thro'
the Zenith and Nadir, are called *Vertical
Circles,* or *Azimuths.* Of these Circles;
that which passes thro' the North and South

Meridian.

Points of the Horizon, is called the *Meridian* ; so that when any Object is upon
the Meridian, it then bears either due South

Azimuth. or due North from us ; and the *Azimuth* of
any

any Object, is an Arch of the Horizon intercepted between the Vertical Circle paffing
thro' it, and either the North or South
Part of the Meridian; which Part is commonly fpecified.

The Meridian paffes thro' the Poles of the
World, as well as thro' the Zenith and Nadir, and therefore is a Secundary both of the
Equinoctial and the Horizon : This Circle divides the Globe into the *Eaftern* and *Weftern*
Hemifpheres; and the *Poles* of it are the
Eaft and *Weft* Points of the *Horizon*. All
the Heavenly Objects are during one half
of their Continuance above the Horizon, in
the Eaftern Hemifphere, and for the other
half in the Weftern; fo that whenever the
Sun arrives upon the upper part of the Meridian, it is then *Noon* or *Mid-day*, which
is the reafon why this Circle is called the
Meridian; and when he comes to the lower
part, it is then *Mid-night*.

The Vertical Circle paffing thro' the Eaft
and Weft Points of the Horizon, is called
the *Prime Vertical*, or *Circle* of *Eaft* and *Prime Ver-*
Weft : fo that when any Object is upon this *tical.*
Circle in the Eaftern Hemifphere, it appears
due Eaft; and if it be in the Weftern Hemifphere, it appears due Weft.

That Degree in the Horizon wherein any
Object rifes or fets from the Eaft or Weft
Points, is called the *Amplitude*; which for *Amplitude*
Rifing is called *Amplitude Ortive*, and

E 2   Oc-

Occasive for Setting; which muft be alfo denominated, whether it be Northerly or Southerly.

It may be obferved, that the *Amplitude* and *Azimuth* are much the fame; the Amplitude fhewing the bearing of any Object when he rifes or fets, from the Eaft or Weft Points of the Horizon; and the Azimuth the bearing of any Object when it is above the Horizon, either from the North or South Points thereof. As for Example, if an Object rifes or fets within 10 Degrees of the Eaft or Weft, fuppofe towards the South, we accordingly fay, its *Amplitude* is 10 Degrees Southerly; but if an Object, that is of any height above the Horizon fhould be in the Vertical Circle paffing thro' the aforementioned Point, we then fay, its *Azimuth* is 80 Degrees from the South, or 100 Degrees from the North, both which Expreffions fignify the fame.

14. All Circles drawn parallel to the Horizon, in the upper Hemifphere, are called *Almacanthers, or Parallels of Altitude:* So that the *Altitude* of any Point in the Heavens, is an Arch of the Vertical Circle paffing thro' that Point, and intercepted betwixt it and the Horizon : and if the Object be upon the Meridian, it is commonly called the *Meridian Altitude.* The Complement of the Altitude, or what it wants of 90 Degrees, is called the *Zenith Diftance.*

Almacanthers.
Altitudes.

Meridian Altitude.

Zenith Diftance.

The

The Horizon (by which we mean the Rational) is reprefented by the upper Surface of the Wooden Frame, wherein the Globes are placed ; upon this Horizon are defcribed feveral Concentrick Circles, the innermoft of which is divided into four times 90 Degrees, beginning at the Eaft and the Weft Points, and counted both ways to the North and South, where they end at 90 Degrees. The Ufe of thefe Divifions is to fhew the Amplitudes of the Sun and Stars at their Rifing and Setting : Alfo in fome convenient Place upon this Horizon, there is commonly noted the Points of the Compafs. Without the aforemention'd Circle, there is drawn the Ecliptick, with its Divifions into Signs and Degrees, and a Circle of Months and Days : The Ufe of thefe two Circles is to ferve as a *Calendar* to fhew the Sun's Place at any time of the Year ; and by that means to find his Place in the *Ecliptick* drawn upon the Globe it felf.

The *Vertical Circles*, and the *Parallels of Altitude*, are fupplied by a thin Plate of Brafs, having a Nut and Screw at one end, to faften it to the Brafs Meridian in the Zenith Point ; which being done, the lower End of it may be put between the Globe it felf and the inner Edge of the Horizon, and fo turned round about to any Point required. The fiducial

E 3 Edge

Edge thereof reprefenting the *Vertical Cir-*
cles, and the *Degrees* upon it, defcribing
Quadrant the Parallels of Altitude.   This thin Plate
of Altitude. is called the *Quadrant of Altitude.*

The Center of the Horizon being the
Place of Obfervation, it is evident, that
this Circle, and all the others belonging to
it, are continually changed which way foever we move; wherefore we may fuppofe
the Horizon, with its Secundaries and Parallels, to inveft the Globe like a *Rete* or
Net; and to be moveable every way round
it. This is very naturally illuftrated by the
Globes; if we move directly North or directly South, the Change made in the Horizon is reprefented, by moving the Brafs
Meridian (keeping the Globe from turning
about its Axis) in the Notches made in
the Wooden Horizon, juft fo much as
we travelled. If our Courfe fhould be due
Eaft or due Weft, the Alterations made thereby
are reprefented, by turning the Globe accordingly about its Axis, the Brafs Meridian
being kept fixed : and if we fteer betwixt
the Meridian and the Eaft or Weft Points,
then we are to turn the Brafs Meridian, and
alfo the Globe about its Axis accordingly.
The Sum of which is, Let the Spectator
be in what Point foever of the Earth's Surface, he'll there gravitate or tend exactly
towards its Center, and imagine himfelf to
be on the higheft Part thereof, (the uneven-
nefs

nefs of the Ground not being here confidered :) wherefore if we turn the Globe in
fuch a manner, as to bring the feveral Progreffive Steps of a Traveller fucceffively to
the Zenith, we fhall then have the fuccef-
five Alterations made in the Horizon, in
every Part of his Journey. This Explication
being well confider'd, will be of help to
Beginners to conceive how the Earth is
every where habitable, and how Paffengers
can travel quite round it : For fince every
thing tends towards the Center of the Earth,
we are to conceive that Point as being the
loweft, and not to carry our Idea of down-
wards any farther : Thofe that are diametrically oppofite to us, being as much upon
the upper part of the Earth as we are ; there
being no fuch thing in Nature, as one Place
being higher than another, but as it is at a
greater diftance from the Center of the
Earth, let it be in what Country foever.

We have now done with all the Circles
of the Sphere, and it may be obferved, that
the *Equinoctial,* the *Ecliptick,* and the *Horizon,* with their Secundaries and Parallels,
are all alike; and altering their Pofition,
may be made to ferve for one another.
Thus, If the *Poles* of the *World* be brought
into the *Zenith* and *Nadir* ; the *Equinoctial* will coincide with the *Horizon,* the
Meridians will be the fame with the *Vertical Circles,* and the Parallels of *Declination*

will be the Parallels of *Altitude.* After the same manner, if shifting the Position we bring the *Ecliptick* to coincide with the *Horizon;* the Circles of *Longitude* will be the *Vertical Circles,* and the Parallels of *Latitude* and *Altitude* will coincide.

The Horizon and the Equator may be either Parallel, Perpendicular, or Oblique to each other.

Parallel Sphere.

15. A *Parallel Sphere* is that Position where the Equator coincides with the Horizon, and consequently the Poles of the World are in the Zenith and Nadir: The Inhabitants of this Sphere (if there be any) are those who live under the Poles of the World.

Right Sphere.

16. A *Right or Direct Sphere* is that Position where the Equator is perpendicular to the Horizon; the Inhabitants whereof are those who live under the E-quinoctial.

Oblique Sphere.

17. An *Oblique Sphere* is when the Equinoctial and the Horizon make Oblique Angles with each other; which every where happens but under the Equator and the Poles.

The Arch of any Parallel or Declination which stands above the Horizon, is called the *Diurnal Arch;* and the remaining part of it, which is below the Horizon, is called the *Nocturnal Arch.*

Diurnal and No-cturnal Arch.

That

That Point of the Eqninoctial which comes to the $\left\{{\text{Eaftern} \atop \text{Weftern}}\right\}$ Part of the Horizon, with any Point in the Heavens, is called the $\left\{{\text{Afcenfion} \atop \text{Defcenfion}}\right\}$ of that Point, counted from the Beginning of ♈; and if it be in a Right Sphere, the Afcenfion or Defcenfion is called Right; but if it be in an Oblique Sphere, it is called Oblique Afcenfion or Defcenfion. So that,

18. The *Right Afcenfion* of the *Sun, Moon*, or any *Star, &c.* is an Arch of the Equator contained betwixt the Beginning of ♈, and that Point of the Equinoctial which rifes with them in a *Right Sphere*; or which comes to the Meridian with them in an Oblique Sphere. *Right Afcenfion.*

19. *Oblique Afcenfion* or *Defcenfion*, is an Arch of the Equinoctial intercepted between the Beginning of ♈, and that *Point* of the *Equator* which rifes or fets with any Point in the Heavens in an Oblique Sphere. *Oblique Afcenfion.*

20. *Afcenfional Difference* is the Difference betwixt the *Right* and Oblique Afcenfion or Defcenfion; and fhews how long the Sun rifes or fets before or after the Hour of Six. *Afcenfional Difference.*

IV. *Of the Divifion of Time.*

The Parts that Time is diftinguifhed into, are *Days, Hours, Weeks, Months* and *Years.*

A

A Day is either Natural or Artificial.

Natural and Artificial Day. A *Natural Day* is the Space of Time elapsed while the Sun goes from any Meridian or Horary Circle, till he arrives to the same again ; or, it is the Time contained from Noon or any particular Hour, to the next Noon or the same Hour again : An *Artificial Day*, is the Time betwixt the Sun's Rising and Setting ; to which is opposed the *Night*, that is, the Time the Sun is hid under the Horizon.

Hours, &c. The *Natural Day* is divided into 24 *Hours*, each Hour into 60 *Minutes*, each Minute into 60 *Seconds*, &c. The *Artificial Days* are always unequal to all the Inhabitants that are not under the Equator, except when the Sun is in the Equinoctial Points ♈ and ♎, which happens (according to our way of reckoning) about the 10th of *March* and 12th of *September* ; at those Times the Sun rises at six and sets at six, to all the Inhabitants of the Earth. These Days are called the *Equinoxes* or Equinoctial Days; *Equinoxes.* the first of which, or when the Sun is in *Vernal and* the first Point of *Aries*, is called the *Vernal* *Autumnal Equinox*, and the latter is called the *Au-* *Equinox.* *tumnal Equinox*. At all other times of the Year, the Days continually lengthen or shorten, and that faster or slower according as the Sun is nearer to or further from the Equinoctial, until he arrives to either of the *Solstitial Points* ♋ or ♑. At those Times the Sun seems to stand still for a few Days, and then

then he begins to return with a flow Motion towards the Equinoctial, still haftening his pace as he comes nearer to it: The Sun enters the Tropicks of ♋ and ♑, about the 10th of our *June* and the 11th of *December*, which Days are fometimes called the *Sol-* ^Solstices. *ftices*; the first of which we call the *Summer* ^Summer ^and Winter *Solstice*, and the latter the *Winter Solstice.* ^Solstice.

All Nations do not begin their Day, and ^The diffe- reckon their Hours alike. In *Great-Britain,* ^rent Begin- *France* and *Spain*, and in moft places in ^ning of the *Europe*, the Day is reckoned to begin at ^Day. Midnight, from whence is counted 12 Hours till Noon, then 12 Hours more till next Midnight, which makes a compleat Day : Yet the *Aftronomers* (in thefe Countries) commonly begin their Day at Noon, and fo reckon 24 Hours till next Noon, and not twice twelve, according to the vulgar Computation.

The *Babylonians* began their Day at Sun- ^Babylonifh rifing, and reckoned 24 Hours till he rofe ^Hours. again : this way of Computation we call the *Babylonifh Hours.* In feveral parts of *Germany* they count their Hours from Sun-fetting, calling the first Hour after the Sun has fet, the first Hour, *&c.* till he fets the next Day, which they call the 24th Hour: thefe are commonly called the *Italian Hours.* ^Italian According to both thefe ways of Computa- ^Hours. tion, their Hours are commonly either a little greater or lefs than the $\frac{1}{24}$ part of a natural Day, in proportion as the Sun rifes or

fets

sets sooner or later in the succeeding Days: They have also this Inconvenience, that their Mid-day and Mid-night happen on different Hours, according to the Seasons of the Year.

The *Jews* and the *Romans* formerly, divided the *Artificial* Days and Nights each into 12 equal parts; these are termed the *Jewish Hours*, and are of different Lengths according to the Seasons of the Year; a Jewish Hour in Summer, being longer than one in Winter, and a Night Hour shorter. This Method of Computation is now in use among the *Turks*, and the Hours are styled the *first Hour, second Hour, &c.* of the Day or Night; so that *Mid-day* always falls upon the sixth Hour of the Day. These Hours are also called *Planetary Hours*; because in every Hour one of the Seven Planets were supposed to preside over the World, and so take it by turns. The first Hour after Sun-rising on Sunday was allotted to the *Sun*; the next to *Venus*; the third to *Mercury*; and the rest in order to the *Moon, Saturn, Jupiter,* and *Mars*. By this means on the first Hour of the next Day, the Moon presided, and so gave the Name to that Day; and so seven Days by this Method had Names given them from the Planets that were supposed to govern on the first Hour.

A *Week* is a System of seven Days, in which each Day is distinguished by a different

Jewish Hours.

Planetary Hours.

A Week.

ferent Name. In moſt Countries, theſe Days are called after the Names of the Seven Planets, as above noted. All Nations that have any Notion of Religion, lay apart one Day in ſeven for publick Worſhip; the Day ſolemnized by *Chriſtians* is Sunday, or the firſt Day of the Week, being that on which our Saviour roſe from the Grave, on which the Apoſtles afterwards uſed more particularly to aſſemble together to perform divine Worſhip. The *Jews* obſerved Saturday, or the ſeventh Day of the Week, for their Sabbath or Day of Reſt, being that appointed in the fourth Commandment under the Law. The *Turks* perform their Religious Ceremonies on Friday.

A *Month* is properly a certain ſpace of *A Month.* Time meaſured by the Moon in its Courſe round the Earth. A *Lunar* Month is either *Periodical* or *Synodical.* A *Periodical* Periodical *Month* is that Space of Time the Moon and Synodical takes to perform her Courſe from one Point Months. of the Ecliptick till ſhe arrives to the ſame again; which is 27 Days and ſome odd Hours: And a *Synodical Month* is the Time betwixt one New Moon and the next New Moon, which is commonly about 29½ Days. But a *Civil Month* is different from theſe, and conſiſts of a certain Number of Days, fewer or more, according to the Laws and Cuſtoms of the Country wherein they are obſerved.

The

The compleateſt Period of Time is a
Year, in which all the Variety of Seaſons
return, and afterwards begin a-new. A
Year is either *Aſtronomical* or *Civil*. An

A Year Sy-dereal and Tropical. *Aſtronomical Year* is either a *Sydereal*,
wherein the Sun departing from a fixed Star,
returns to it again ; or, *Tropical*, which is
the ſpace of Time the Sun takes to perform
its Courſe from any Point of the Ecliptick
till he returns to it again.

A *Tropical Year* conſiſts of 365 Days,
5 Hours, and 49 Minutes ; this is the Time
in which all the Seaſons compleatly return,
which is a ſmall matter leſs than a Sydereal
Year.

The *Civil Year* is the ſame with the *Po-litical* eſtabliſhed by the Laws of a Coun
try, and is either moveable or immoveable.
The moveable Year conſiſts of 365 Days,
being leſs than the Tropical Year by almoſt

Egyptian Year. ſix Hours ; and is called the *Egyptian Year*,
becauſe obſerved in that Country.

The *Romans* divided the Year into 12 Ka-lendar Months, to which they gave particu-lar Names, and are ſtill retained by moſt of
the *European* Nations, *viz.* *January*, *Fe-bruary*, *March*, *April*, *May*, *June*, *July*,
Auguſt, *September*, *October*, *November*,
and *December*. The Number of Days in
each Month may be known by the follow-ing Verſes:

Thirty

Thirty Days hath September,
April, June, *and* November ;
February *hath Twenty Eight alone*,
And all the reft have Thirty One.

The Year is alfo divided into four Quarters
or Seafons, *viz.* The *Spring, Summer, Au-*
tumn and *Winter.* Thefe Quarters are properly made when the Sun enters into the
Equinoctial and Solftitial Points of the
Ecliptick : But in Civil Ufes they are dif-
ferently reckon'd according to the Cuftoms
of feveral Countries. In *England* we commonly reckon the firft Day of *January* to
be the firft in the Year, which is therefore
vulgarly called *Newyear's-Day* ; but in Po-
litical and Ecclefiaftical Affairs, the Year is
reckoned to commence on *Lady Day*,
which is the 25th of *March*; and from
thence to *Midfummer Day*, which is the
24th of *June*, is reckoned the firft Quarter ; from *Midfummer-Day* to *Michaelmas-*
Day, which is the 29th of *September*, is
the fecond Quarter ; the third Quarter
is reckon'd from *Michaelmas-Day* to
Chriftmas-Day, which is the 25th of *De-*
cember ; and from *Chriftmas-Day* to *La-*
dy-Day, is reckoned the laft Quarter in
the Year. In common Affairs, a Quarter is reckoned from a certain Day to the
fame in the fourth Month following.
Sometimes a Month is reckoned four
    I   Weeks

Weeks or 28 Days, and so a Quarter 12 Weeks. To all the Inhabitants in the {Northern} {Southern} Hemisphere, their *Midsummer* is properly when the Sun is in the Tropick of {Cancer} {Capricorn} and their *Midwinter* at the opposite Time of the Year. But those who live under the Equinoctial, have two Winters, *viz.* when the Sun is in either Tropick; tho' indeed properly, there is no Season that may be called Winter, in those parts of the World.

The *Egyptian* Year of 365 Days, being less than the true Solar Year, by almost six Hours, it follows that four such Years are less than four Solar Years, by a whole Day; and therefore in 365 times four Years, that is, in 1460 Years, the beginning of the Years move through all the Seasons. To remedy this Inconveniency, *Julius Cæsar* (considering that the six Hours which remain at the end of every Year, will in four Years make a Natural Day) ordered that every fourth Year should have an intercalary Day, which therefore consists of 366 Days; the Day added was put in the Month of *February*, by postponing St. *Matthias*'s Day, which in common Years falls on the 24th, to the 25th of the said Month: all the fixed Feasts in the Year from thenceforwards falling a Week-day later than otherwise they would.   Ac-

According to the *Roman* way of reckon- ing, the 24th of *February* was the fixth of the Kalends of *March*, and it was ordered that for this Year there mould be two fixths, or that the fixth of the Kalends of *March* mould be twice repeated; upon which account the Year was called *Biffextile*, which we now call the *Leap-year*.

To find whether the Year of our Lord be Leap Year, or the first, fecond, or third after; Divide it by four, and the remainder, if there be any, mews how many Years it is after Leap Year; but if there be no Remainder, then that Year is Leap Year : Or, you may omit the Hundreds and Scores, and divide the refidue by 4. *Examp.* 1731, omitting the Hundreds and the Twenty, I divide the refidue 11 by 4, and the remainder 3 mews it to be the third after Leap-Year.

This Method of reckoning the Year, *viz.* making the common Year to confift of 365 Days, and every fourth Year to have 366 Days, is now ufed in *Great-Britain* and *Ireland,* and fome of the Northern Parts of *Europe,* and is called the *Julian Account* or the *Old Style.* But the Time appointed by *Julius Cæfar* for the Length of a Solar Year is too much; for the Sun finishes his Courfe in the Ecliptick in 365 Days, 5 Hours, and 49 Minutes, which is 11 Minutes lefs than the Civil Year; and therefore he again begins his Circuit 11 Minutes before the

F   Civil

Civil Year is ended : And so much being
gained every Year, amounts in 131 Years
to a whole Day. So that if the Sun in any
Year entered the Equinox upon the 20th of
March at Noon, after the Space of 131 Years,
he'll enter the same Point on the same Hour
of the 19th of *March.* And therefore the
Equinoxes will not always fall on the same
Day of the Month, but by degrees will
move towards the beginning of the Year.

At the Time of the *Council of Nice,*
(when the Terms were settled for Observ-
ing of *Easter*) the *Vernal Equinox* fell upon the 21st of *March ;* but by its falling
backwards 11 Minutes every Year, it was
found that in *Anno* 1582, when the Kalendar was corrected, the Sun entered the Equinoctial Circle on the 11th of *March,* having
departed ten whole Days from its former
Place in the Year : And therefore Pope *Gre-*
gory the XIIIth designing to place the Equinoxes in their former situation with respect
to the Year, took these ten Days out of the
Kalendar, and order'd that the Eleventh of
March should be reckon'd as the Twenty-
first : And to prevent the Seasons of the
Year from going backwards for the future, he
ordered that every Hundredth Year, which
in the *Julian* Form was to be a *Bissextile,*
should be a common Year, and consist only
of 365 Days ; but that being too much,
every fourth Hundred was to remain *Bissex-*
tile. This Form of reckoning being esta-
blished

blifhed by the Authority of Pope *Gre-*
gory XIII. is called the *Gregorian Account,* Gregorian
or the *New Style*; and is obferved in all New Style.
the Countries where the Authority of the
Pope is acknowledged, and likewife, by feveral Nations of the Reformed Religion. There
being now above a Hundred Years paft, fince
this Reformation was made in the Kalendar,
the *Gregorian* Account has accordingly got
before the *Julian* one Day more than it
was at the time of its Inftitution, the Difference between thefe two Accounts being
now eleven Days; fo that the firft Day of
any Month, according to our way of reckoning, is the 12th of the fame Month according to the New Style.

I fhall conclude this Section with a brief
Account of the Atmofphere.

The *Atmofphere* is that thin Body of Air Atmo-
which furrounds the Earth, in which the fphere.
Clouds hover, and by which in their defcent
they are broke into Drops of Rain ; which
fometimes, according to the warmth or cold-
nefs of the Air, are froze into *Snow*, or
Hailftones. Thunder and *Lightning* are
alfo made in the *Atmofphere* ; and Wind
is nothing elfe but a Percuffion of the Air,
occafion'd by its different Denfity in different
Places. The Benefits we receive from the
Atmofphere are innumerable ; without Air
no earthly Creature could live, as is plainly
proved by Experiments made by the *Air-*

Pump; and the Wholefomenefs of a Climate chiefly depends upon that of its Air: If there was no Atmofphere to reflect the Rays of the Sun, no part of the Heavens would be lucid and bright, but that wherein the Sun was placed; and if a Spectator fhould turn his Back towards the Sun, he would immediately perceive it to be quite dark, and the leaft Stars would be feen fhining as they do in the cleareft Night; and the Sun immediately before his fetting would fhine as brisk as at Noon, but in a Moment, as foon as he got below the Horizon, the whole Hemifphere of the Earth would be involved in as great a Darknefs, as if it were Midnight.

But by means of the Atmofphere, it happens that while the Sun is above the Horizon, the whole Face of the Heavens is ftrongly illuminated by its Rays, fo as to obfcure the faint Light of the Stars and render them Invifible; and after Sun-fetting, though we receive no direct Light from him, yet we enjoy its reflected Light for fome Time: for the Atmofphere being higher than we are, is a longer Time before it is withdrawn from the Sun, (as if a Man was to run up to the Top of a Steeple, he may fee the Sun after it had been fet to thofe at the Bottom.) The Rays which the Atmofphere receives from the Sun, after he is withdrawn from our Sight, are by Refraction
faintly

faintly tranfmitted to us ; until the Sun having got about 18 Degrees below the Horizon, he no longer enlightens our Atmofphere, and then all that Part thereof which is over us becomes dark. After the fame manner, in the Morning, when the Sun comes within 18 Degrees of our Horizon, he again begins to enlighten the Atmofphere, and fo more and more by degrees, until he rifes and makes full Day. This fmall Illumination of the Atmofphere, and State of the Heavens between Day and Night, is called the *Twilight*, or the *Crepufculum*. *Twilight, or the Crepufculum.*

The Duration of Twilight is different in different Climates, and in the fame Place at different Times of the Year. The Beginning or Ending of Twilight being accurately given, we may from thence eafily find the Height of the Atmofphere, which is not always the fame. The mean Height of the Atmofphere is computed to be about 40 Miles; but it is probable, the Air may expand it felf a great deal further, there being properly no other Limits to it, as we can conceive, but as it continually decreafes in Denfity the farther remote it is from the Earth, in a certain Ratio ; which at laft, as to our Conception, muft terminate.

F 3   SECT.

❦❦❦❦❦❦❦❦❦❦❦

S E C T. II.

GEOGRAPHICAL DEFINITIONS:
*Of the Situations of Places upon
the Earth ; of the different Situations of its Inhabitants ; of Zones
and Climates.*

THE Situation of Places upon the Earth
are determined by their Latitude
and Longitude.

Latitude.   1. The *Latitude* of any Place (upon the
Earth) is its neareſt Diſtance, either North
or South, from the Equator; and if the
Place be in the $\begin{Bmatrix} \text{Northern} \\ \text{Southern} \end{Bmatrix}$ Hemiſphere,

it is accordingly called $\begin{Bmatrix} North \\ South \end{Bmatrix}$ *Latitude* ;

and is meaſured by an Arch of the Meridian
paſſing thro' the Zenith of the ſaid Place,
and intercepted betwixt it and the Equator.
And all Places that lie on the ſame Side,
and at the ſame Diſtance from the Equator,
are ſaid to be in the ſame Parallel of Latitude . The Parallels of Latitude in *Geogra-*
                  ɪ   *phy*

phy being the fame with the Parallels of Declination in *Aſtronomy*.

From this Definition ariſe the following Corollaries.

(1.) *That no Place can have above* 90 *Degrees of Latitude, either North or South.*

(2.) *Thoſe Places that lie under the Equinoctial (or thro' which the Equator paſſes) have no Latitude, it being from thence that the Calculation of Latitude is counted. And thoſe Places that lie under the Poles have the greateſt Latitude, thoſe Points being at the greateſt diſtance from the Equator.*

(3.) *The Latitude of any Place is always equal to the Elevation of the Pole in the ſame Place, above the Horizon; and is therefore often expreſſed by the Pole's Height, or Elevation of the Pole: The reaſon of which is, becauſe from the Equator to the Pole, there is always the Diſtance of* 90 *Degrees, and from the Zenith to the Horizon the ſame Number of Degrees, each of theſe including the Diſtance from the Zenith to the Pole. That Diſtance therefore being taken away from both, will leave the Diſtance from the Zenith to the Equator, (which is the Latitude) equal to the Diſtance from the Pole to the Horizon.*

(4.) *The Elevation of the Equator in any Place, is always equal to the Complement of the Latitude of the ſame Place.*

(5.) *A Ship sailing directly* $\left\{\begin{array}{l}towards\\ from\end{array}\right\}$

the Equator, $\left\{\begin{array}{l}Lessens\\ Augments\end{array}\right\}$ *her Latitude,*

(or $\left\{\begin{array}{l}Depresses\\ Raises\end{array}\right\}$ *the Pole) just so much*

as is her Distance sailed.

Difference of Latitude.   2. *Difference of Latitude* is the neareft Diftance betwixt any two Parallels of Latitude, fhewing how far the one is to the Northward or Southward of the other; which can never exceed 180 Degrees. And when the two Places are in the fame Hemifphere (or on the fame Side of the Equator) the leffer Latitude fubtracted from the greater, and when they are on different Sides of the Equator, the two Latitudes added, gives the difference of Latitude.

Longitude.   3. The *Longitude* of any Place (upon the Earth) is an Arch of the Equator, contained betwixt the Meridian of the given Place, and fome fixed or known Meridian: or, it is equal to the Angle formed by the two Meridians; which properly can never exceed 180 Degrees, tho fometimes the Longitude is counted Eafterly quite round the Globe.

Since the Meridians are all moveable, and not one that can be fixed in the Heavens, (as the Equinoctial Circle is fixed, from whence the Latitudes of all Places are determined to be fo much either North

or

or South) the Longitudes of Places cannot fo well be fixed from any one Meridian; but every Geographer is at his liberty
to make which he pleafes his firft Meridian,
from whence to calculate the Longitudes
of other Places. Hence it is, that the Geographers of different Nations, reckon their
Longitudes from different Meridians, commonly chufing the Meridian paffing thro'
the Metropolis of their own Country for
their firft: Thus, the *Englifh* Geographers
generally make the *Meridian* of *London*
to be their firft; the *French*, that of *Paris*;
and [the *Dutch*, that of *Amfterdam, &c.*
And Mariners generally reckon their Longitude from the laft known Land they faw.
This arbitrary way of reckoning the Longitude from different Places, makes it neceffary, whenever we exprefs the Longitude
of any Place, that the Place from whence
it is counted be alfo expreffed.

From the preceding Definitions arife the
following Corollaries.
    1. *If a Body fhould fteer direɛtly North,*
or South; quite round the Globe, he'll conti-
nually change his Latitude; and pafs thro'
the two Poles of the World, without de-
viating the leaft from the Meridian of the
Place he departed from; and confequently
at his return, will not differ in his Account
of Time from the People refiding in the
faid Place.   2,

2. *If a Body should steer round the Globe, either due East or due West, he'll continually change his Longitude, but will go quite round without altering his Latitude; and if his Course should be due East, he'll gain a Day compleatly in his reckoning, or reckon one Day more than the Inhabitants of the Place from whence he departed: and if his Course had been West, he would have lost one Day, or reckoned one less.*

The Reason of which is evident : For, admitting our Traveller steers due East so many Miles in one Day, as to make his Difference of Longitude equivalent to a quarter of an Hour of Time; it is evident, that the next Day the Sun will rise to him a quarter of an Hour sooner, than to the Inhabitants of the Place from whence he departed: and so daily, in proportion to the rate he travels, which in going quite round will make up one Natural Day. In like manner, if he steers due West after the same rate, he'll lengthen each Day a Quarter of an Hour, and consequently the Sun will rise to him so much later every Day; by which means, in going quite round, he'll lose one Day compleat in his reckoning. From whence it follows,

3. *If two Bodies should set out from the same Place, one steering East, and the other West, and so continue their Courses quite round, until they arrive at the Place from*
whence

whence they set out, they'll differ two Days in their Reckoning at the Time of their return.

4. If a Body should steer upon an Oblique Course (or any where betwixt the Meridian and the East or West Points) he'll continually change both Latitude and Longitude, and that more or less, according to the Course he steers; and if he should go quite round the Globe, he'll differ in his Account of Time, as by the 2d Corol.

5. The People residing in the Eastermost of any two Places, will reckon their Time so much the sooner than those who live in the other Place, according to the Difference of Longitude betwixt the two Places, allowing 1 Hour for every 15 Degrees, &c. and the contrary.

II. Of Zones and Climates, &c.

4. *Zones* are large Tracts of the Surface of the Earth, distinguished by the Tropicks and Polar Circles; being five in Number, *viz.* one *Torrid,* two *Temperate,* and two *Frigid.* {Zones Torrid, Temperate, and Frigid.}

The *Torrid* or *Burning Zone* is all the Space comprehended between the two Tropicks; the Antients imagined this Tract of the Earth to be uninhabitable, because of the excessive Heat, it being so near the Sun. All the Inhabitants of the Torrid Zone have

have the Sun in their Zenith, or exactly over their Heads, twice in every Year; excepting those who live exactly under the two Tropicks, where the Sun comes to their Zenith only once in a Year.

The two *Temperate Zones* lie on either Side of the Globe, between the Tropicks and the Polar Circles.

The two *Frigid Zones* are those Spaces upon the Globe that are included within the two Polar Circles.

The Inhabitants of the Earth are also distinguished by the Diversity of their *Shadows*. Those who live in the Torrid Zone, are called *Amphiscians*; because their Noon-Shadow is cast different Ways, according as the Sun is to the Northward or Southward of their Zenith: But when the Sun is in their Zenith, they are called *Ascians*.

Amphisci-ans.

Ascians.

The Inhabitants of the Temperate Zones, are called *Heteroscians*, because their Noon-Shadow is always cast the same way: But those who live under the Tropicks are called *Ascians Heteroscians.* Those who live in the Frigid Zones are called *Periscians*, because sometimes their Shadow is cast round about them.

Heterosci-ans.

Ascians Heterosci-ans.
Periscians.

These hard Names are only *Greek* Words, importing how the Sun casts the Shadow of the several Inhabitants of the Earth; which would be a too trifling distinction to be made here, was it not for the sake of complying with Custom.   The

The Inhabitants of the Earth, are alſo diſtinguiſhed into three Sorts, in reſpect to their relative Situation to one another; and theſe are called, the *Periæci, Antæci,* and *Antipodes.*

5. The *Periæci* are thoſe who live under oppoſite Points of the ſame Parallel of Latitude.   They have their Seaſons of the Year at the ſame Time, and their Days and Nights always of the ſame Length with one another; but the one's *Noon* is the other's *Midnight:* and when the Sun is in the Equinoctial, he riſes with the one, when he ſets with the other.   Thoſe who live under the Poles have no *Periæci.* *Periæci*

6. The *Antæci* live under the ſame Meridian, and in the ſame Latitude, but on different Sides of the Equator; their Seaſons of the Year are contrary, and the Days of the one are equal to the Nights of the other: but the Hour of the Day and Night is the ſame with both; and when the Sun is in the Equinoctial, he riſes and ſets to both exactly at the ſame time.   Thoſe who live under the Equator have no *Antæci.* *Antæci*

7. The *Antipodes* are thoſe who live diametrically oppoſite to one another, ſtanding, as it were, exactly feet to feet: Their Days and Nights, Summer and Winter, are at direct contrary Times. *Antipodes.*

The Surface of the Earth is by ſome diſtinguiſhed into *Climates.*

8. A

8. A *Climate* is a Tract of the Surface of the Earth, included between two such Parallels of Latitude, that the Length of the longest Day in the one, exceeds that in the other by half an Hour.

The whole Surface of the Earth, is considered as being divided into 60 Climates, *viz.* from the Equator to each of the Polar Circles 24, arising from the Difference of $\frac{1}{2}$ Hour in the length of their longest Days; and from the Polar Circles to the Poles themselves are six, arising from the difference of an entire Month; the Sun being seen in the first of these, a whole Month without setting; in the second, two; and in the third, three Months, &c. These Climates continually decrease in Breadth the farther they are from the Equator. How they are framed, *viz.* the Parallel of Latitude in which they end (that being likewise the Breadth of the next) with the respective Breadth of each of them, is shewed in the following Table.

A

A TABLE *of the* CLIMATES.

CLIMATES *between the* Equator *and the* Polar Circles.

Climates.	Longest Day.	Latitude D.	M.	Breadth D.	M.	Climates.	Longest Day.	Latitude D.	M.	Breadth D.	M.
I	12½	8	25	8	25	13	18½	59	58	1	29
2	13	16	25	8	00	14	19	61	18	1	20
3	13½	23	50	7	25	15	19½	62	25	1	07
4	14	30	20	6	30	16	20	63	22	0	57
5	14½	36	28	6	08	17	20½	64	06	0	44
6	15	41	22	4	54	18	21	64	49	0	43
7	15½	45	29	4	07	19	21½	65	21	0	32
8	16	49	01	3	32	20	22	65	47	0	26
9	16½	51	58	2	57	21	22½	66	06	0	19
10	17	54	27	2	29	22	23	66	20	0	14
11	17½	56	37	2	10	23	23½	66	28	0	08
12	18	58	29	1	52	24	24	66	31	0	03

CLIMATES *between the* Polar Circles *and the* Poles.

Length of Days. Months.	Latitude. D.	M.	Length of Days. Months.	Latitude. D.	M.
1	67	21	4	78	30
2	69	48	5	84	05
3	73	37	6	90	00

III. *Of*

III. *Of the Poetical Rising and Setting of the Stars.*

The Antient Poets make frequent mention of the Stars Rising and Setting, either *Cosmical,* *Cosmically,* *Achronically,* or *Heliacally;* *Achronical* whence these Distinctions are called *Po-* *and Helia-* *etical.*
cal rising *and setting.*

A Star is said to *Rise* or *Set Cosmically,* when it rises or sets at Sun-rising; and when it rises or sets at Sun-setting, it is said to rise or set *Achronically.* A Star *rises He-liacally* when first it becomes visible, after it had been so near the Sun as to be hid by the Splendor of his Rays : And a Star is said to *set Heliacally,* when 'tis first immersed, or hid by the Sun's Rays.

The *Fixed Stars,* and the three superior Planets, *Mars, Jupiter,* and *Saturn,* rise Heliacally in the Morning : but the *Moon* rises Heliacally in the Evening; because the Sun is swifter than the superior Planets, and slower than the Moon.

IV. *Of the Surface of the Earth, considered as it is composed of Land and Water.*

The Earth consists naturally of two Parts, Land and Water; and therefore it is called the *Terraqueous Globe.* Each of these Elements are subdivided into
various

various Forms and Parts, which accordingly
are diftinguifhed by different Names.

I. *Of the Land.*

The Land is diftinguifhed into *Continents,
Iflands, Peninfula's, Ifthmus's, Promonto-
ries, Mountains,* or *Coafts.*

9. A *Continent* is a large Quantity of Land, in which many great Countries are joined together, without being feparated from each other by the Sea: Such are *Europe, Afia, Africa,* and the Vaft Continent of *America*; which four are the principal Divifions of the Earth. A Continent is fometimes called the *Main Land.* <small>Continent.</small> <small>MainLand.</small>

10. An *Ifland* is a Country, or Portion of Land environed round with Water: Such are *Great Britain,* and *Ireland*; *Sardinia, Sicily, &c.* in the Mediterranean Sea; the *Ifles* of *Wight, Anglefey, &c.* near *England.* Alfo a fmall part of dry Land in the midft of a River, is called an Ifland. Sometimes a large Ifland, when compared to a leffer, is called the Continent; as if we compare the *Ifle* of *Wight* to *England,* the latter may be properly called the Continent.

11. A *Peninfula* is a Part of Land almoft environed with Water, fave one narrow Neck adjoining it to the Continent; Or, which is almoft an Ifland. Such is *Denmark* <small>Peninfula.</small>

G   *mark*

mark joining to *Germany*; alſo *Africa* is properly a large Peninſula joining to *Aſia*.

Iſthmus.   12. An *Iſthmus* is a narrow Neck of Land joining a Peninſula to the Continent; as the *Iſthmus* of *Sues*, which joins *Africa* to *Aſia*; that of *Panama* joining North and South *America*, *&c.*

Promon- *tory.*   13. A *Promontory* is a high Part of Land ſtretching out into the Sea; and is often called a *Cape*, or *Headland*: Such is the *Cape of Good Hope*, in the South of *Africa*; *Cape Finiſtre* on the Weſt of *Spain*; alſo the *Lizard Point*, and the *Land's End*, are two Capes or Headlands on the Weſt of

Mountain *England*. A *Mountain* is a high Part of Land in the Midſt of a Country, over-topping the adjacent Parts.

A Coaſt or *Shore.*   14. A *Coaſt* or *Shore* is that Part of Land which borders upon the Sea, whether it be in Iſlands or a Continent : And that Part of the Land which is far diſtant from the Sea,

Inland. is called the *Inland Country*. Theſe are the uſual Diſtinctions of the Land.

The Water is diſtinguiſhed into *Oceans*, *Seas*, *Lakes*, *Gulfs*, *Streights*, and *Rivers*.

The Ocean *or Main* *Sea.*   15. The *Ocean* or *Main-Sea*, is a vaſt ſpreading Collection of Water, not divided or ſeparated by Lands running between : Such is the *Atlantick* or *Weſtern Ocean*, between *Europe* and *America*; the Pacifick Ocean or *South-Sea*, *&c.*

                                        Note,

Note, Thofe Parts of the Ocean, which border upon the Land, are called by various Names, according to thofe of the adjacent Countries; · as the *Britifh Sea ;* the *Irifh Sea ;* the *French* and *Spanifh Sea.*

16. A *Lake* is a Colleftion of deep ftand- *A Lake.* ing Water, enclofed all round with Land, and not having any vifible and open Communication with the Sea: But when this Lake is very large, it is commonly called a Sea, as the *Cafpian Sea* in *Afia, &c.*

17. A *Gulf* is a Part of the Sea almoft *A Gulf.* encompaffed with Land, or, that which runs up a great way into the Land ; as the *Gulf of Venice, &c.* But if it be very large, 'tis rather called an *Inland Sea;* as the *Baltick Sea,* the *Mediterranean Sea,* the *Red Sea,* or the *Arabian Gulf, &c.* And a fmall Part of the Sea thus environed with Land, is ufually called a *Bay.* If it be but a very fmall Part, or as it were, a fmall Arm of the Sea that runs but a few Miles between *Creek or* the Land, it is called a *Creek* or *Haven. Haven.*

18. A *Streight* is a narrow Paffage lying *A Streight.* between two Shores, whereby two Seas are joined together ; as the *Streights of Dover,* between the *Britifh* Channel, and the *German* Sea; the *Streights of Gibraltar,* between the Atlantick and the Mediterranean Sea. The Mediterranean it felf is alfo fometimes called the *Streights.*

Thefe are all the neceffary Terms commonly ufed in *Geography.* The Names of
G 2   the

the feveral Countries, Seas, and all the principal Divifions of the Earth, the Reader will
find expreffed upon the Terreftrial Globes.
To give a tolerable Account of the Produce
of each Country, the Genius of the People,
their Political Inftitutions, *&c.* is properly a
particular Subject of it felf ; and quite foreign
to our Defign. We fhall next proceed to
the Ufe of the Globes; but firft it may not
be amifs to give a fhort *Review* of their Ap-
purtances.

Thofe Circles of the Sphere that are *fixed*
are (as has been already faid) drawn upon
the *Globes* themfelves; thofe that are *moveable*, are fupplied by the *Brafs Meridian*,
the *Wooden Horizon*, and the *Quadrant of
Altitude*.

Brafs Me-
ridian.   1. That Side of the *Brazen Meridian*,
which is divided into Degrees, reprefents
the *true Meridian*; this Side is commonly
turned towards the Eaft, and 'tis ufual to
place the Globe fo before you, that the
North be to the Right-Hand, and the South
to the Left. The Meridian is divided into
4 Quadrants, each being 90 Degrees,. two
of which are numbered from that Part of
the Equinoctial which is above the Horizon,
towards each of the Poles ; the other two
Quadrants are numbered from the Poles
towards the Equator. The reafon why two
Quadrants of the Meridian are numbered
from the Equator, and the other two from
the

the Poles, is becaufe the former of thefe
two ferve to fhew the Diftance of any Point
on the Globe from the Equator, and the
other to elevate the Globe to the Latitude
of the Place.

2. The upper Side of the Wooden *Wooden Horizon*
Frame, called the Wooden Horizon, reprefents the true Horizon; the Circles
drawn upon this Plane have been already
defcribed; we may obferve, that the firft
Point of ♈ is the Eaft, and the oppofite,
being the firft Point of ♎ is the Weft, the
Meridian paffing thro' the North and South
Points.

3. The *Quadrant of Altitude* is a flexible *Quadrant of Altitude.*
Plate of thin Brafs, having a Nut and Screw
at one End, to be faftened to the Meridian
of either Globe, as occafion requires. The
Edge of this Quadrant, which has the Graduations upon it, called the fiducial Edge,
is that which is always meant whenever we
make mention of the Quadrant of Altitude.

4. The *Horary* or *Hour Circle* is divided *Hour Circle.*
into twice twelve Hours; the two XII's
coinciding with the Meridian: the upper-
moft XII is that at *Noon,* and the lower-
moft towards the Horizon is XII at *Night.*
The Hours on the *Eaft* Side of the Meridian are the *Morning Hours,* and thofe on
the *Weft* Side the *Hours after Noon.* The
Axis of the Globe carries round the *Hand*
or *Index* which points the Hour, and it is
the Center of the Hour Circle. The

The Things above described are common to both Globes ; but there are some others which are peculiar, or proper to one sort of Globe. The two *Colures,* and the *Circles* of *Latitude,* from the Ecliptick, belong only to the *Celestial Globe* ; also the Ecliptick it self does properly belong only to this Globe, tho' it is always drawn on the Terrestrial, for the sake of those that might not have the other Globe by them. The Equinoctial on the Celestial Globe is always numbered into 360 Degrees, beginning at the Equinoctial Point ♈ ; but on the Terrestrial, it is arbitrary where these Numbers commence, according to the Meridian of what Place you intend for your first ; and the Degrees may be counted either quite round to 360, or both Ways, till they meet in the opposite Part of the Meridian at 180.

S E C T.

᥆᥆᥆ ᥆᥆᥆᥆᥆᥆᥆᥆᥆᥆᥆᥆᥆᥆᥆᥆᥆᥆

S E C T. III.

The U S E *of the* GLOBES.

PROBLEM I. *To find the Latitude and Longitude of any given Place upon the Globe; and on the contrary, the Latitude and Longitude being given to find the Place.*

1. TURN the Globe round its Axis, till the given Place lie exactly under the (Eaſtern Side of the Braſs) Meridian ; then that Degree upon the Meridian, which is directly over it, is the *Latitude:* which is accordingly North or South, as it lies in the Northern or Southern Hemiſphere. The Globe remaining in the ſame Poſition.

That Degree upon the Equator which is cut by the Brazen Meridian, is the *Longitude* required, from the firſt Meridian upon the Globe. If the Longitude is counted both ways from the firſt Meridian upon the Globe, then we are to conſider, whether the given Place lies Eaſterly or Weſterly from the firſt Meridian, and the Longitude muſt be expreſſed accordingly.

The

The *Latitudes* of the following Places ; and upon a Globe where the Longitude is reckoned both ways from the Meridian of *London* their *Longitudes*, will be found as follow.

	Latitude.	Longitude.
	Deg.	Deg.
Rome ———	$41\frac{3}{4}$ North.	13 East.
Paris ———	$48\frac{3}{4}$ N.	$2\frac{1}{4}$ E.
Mexico ———	20 N.	102 W.
Cape Horn ———	58 S.	80 W.

2. *The Latitude and Longitude being given, to find the Place.*

Seek for the given Longitude in the E-quator, and bring that Point to the Meri-dian ; then count from the Equator on the Meridian, the Degree of Latitude given, towards the Arctick or Antarctick Pole, ac-cording as the Latitude is Northerly or Southerly ; and under that Degree of Lati-tude lies the *Place* required.

PROB. II. *To find the Difference of Lati-tude betwixt any two given Places.*

Bring each of the Places proposed suc-cessively to the Meridian, and observe where they intersect it ; then the Number of De-grees upon the Meridian, contained between
the

the two Interfections, will be the *Difference
of Latitude* required. Or, if the Places
propofed are on the fame Side of the Equator, having firft found their Latitudes, fub-
tract the leffer from the greater ; but if they
are on the contrary Sides of the Equator,
add them both together ; and the Difference
in the firft Cafe, and the Sum in the latter,
will be the Difference of Latitude required.

Thus, the Difference of Latitude betwixt
London and *Rome* will be found to be
9 ¾ Degrees ; betwixt *Paris* and *Cape Bona
Efperance* 83 Degrees.

PROB. III. *To find the Difference of Longitude betwixt any two given Places.*

Bring each of the given Places fucceffively
to the Meridian, and fee where the Meridian
cuts the Equator each Time; the Number
of Degrees contained betwixt thofe two
Points, if it be lefs than 180 Degrees, other-
wife the Remainder to 360 Degrees will
be the Difference of *Longitude* required.
Or,

Having brought one of the given Places
to the Meridian, bring the Index of the
Hour Circle to 12 o'Clock ; then having
brought the other Place to the Meridian,
the Number of Hours contained between
the Place the Index was firft fet at, and the
Place where it now points, is the Difference
of

of Longitude in Time betwixt the two Places.

Thus, the Difference of Longitude betwixt *Rome* and *Constantinople*, will be found to be 19 Degrees, or 1 Hour and a Quarter; betwixt *Mexico* and *Pekin* in *China* 140 Degrees, or $9\frac{1}{3}$ Hours.

PROB. IV. *Any Place being given, to find all those Places that are in the same Latitude with the said Place.*

The Latitude of the given Place being marked upon the Meridian, turn the Globe round its Axis, and all those Places that pass under the said Mark, are in the same Latitude with the given Place, and have their Days and Nights of equal Lengths. And when any Place is brought to the Meridian, all the Inhabitants, that lie under the upper Semicircle of it, have their Noon or Midday at the same Point of absolute Time exactly.

PROB. V. *The Day of the Month being given; to find the Sun's Place in the Ecliptick, and his Declination.*

1. *To find the Sun's Place :* Look for the Day of the Month given in the Kalendar of Months upon the Horizon, and right against it you'll find that Sign and Degree of the
Eclip-

Ecliptick which the Sun is in : The Sun's
Place being thus found, look for the same
in the Ecliptick Line which is drawn upon
the Globe, and bring that Point to the Meridian; then that Degree of the Meridian
which is directly over the Sun's Place is the
Declination required, which is accordingly either North or South, as the Sun is in
the Northern or Southern Signs. Thus

	Sun's Place.		Declination.		
	Deg.	Min.	Deg.	Min.	
April 12	♉ 3	00	12	32	N.
July 20	♋ 7	51	18	20	N.
October 15	♏ 2	49	12	28	S.
January 9	♒ 0	17	20	07	S.

PROB. VI. *To rectify the Globes for the
Latitude, Zenith, and the Sun's Place.*

1. *For the Latitude:* If the Place be in
the Northern Hemisphere, raise the Arctick
Pole above the Horizon; but for a South
Latitude, you must raise the Antarctick;
then move the Meridian up and down in
the Notches, until the Degrees of the Latitude counted upon the Meridian below the
Pole, cuts the Horizon, and then the Globe
is adjusted to the Latitude.

2. *To rectify the Globe for the Zenith:*
Having elevated the Globe according to the
Latitude, count the Degrees thereof upon
the

the Meridian from the Equator towards the
elevated Pole, and that Point will be the Zenith or the Vertex of the Place: To this Point
of the Meridian faſten the Quadrant of Altitude, ſo that the graduated Edge thereof
may be joined to the ſaid Point.

3. Bring the Sun's Place in the Ecliptick
to the Meridian, and then ſet the Hour-
Index to XII at Noon, and the Globe will
be rectified *to the Sun's Place.*

If you have a little Mariner's Compaſs,
the Meridian of the Globe may be eaſily ſet
to the Meridian of the Place.

PROB. VII. *To find the Diſtance between
any two given Places upon the Globe,
and to find all thoſe Places upon the
Globe that are at the ſame Diſtance from
a given Place.*

Lay the Quadrant of Altitude over both
the Places, and the Number of Degrees
intercepted between them being reduced
into Miles, will be the Diſtance required:
Or, you may take the Diſtance betwixt the
two Places with a Pair of Compaſſes, and
applying that Extent to the Equator, you'll
have the Degrees of Diſtance as before.
Note, A *Geographical Mile* is the $\frac{1}{60}$th
part of a Degree; wherefore if you multiply the Number of Degrees by 60, the Pro-
duct will be the Number of Geographical
Miles

Miles of Diftance fought ; but to reduce the fame into *Englifh* Miles, you muft multiply by 70, becaufe about 70 *Englifh* Miles make a Degree of a great Circle upon the Superficies of the Earth.

Thus, the Diftance betwixt *London* and *Rome* will be found to be about 13 Degrees, which is 780 Geographical Miles.

If you rectify the Globe for the Latitude and Zenith of any given Place, and bring the faid Place to the Meridian ; then turning the Quadrant of Altitude about, all thofe Places that are cut by the fame Point of it, are at the fame Diftance from the given Place.

P R O B. VIII. *To find the Angle of Pofition of Places ; or, the Angle formed by the Meridian of one Place, and a great Circle paffing through both the Places.*

Having rectified the Globe for the Latitude and Zenith of one of the given Places, bring the faid Place to the Meridian, then turn the Quadrant of Altitude about until the fiducial Edge thereof cuts the other Place, and the Number of Degrees upon the Horizon contained between the faid Edge and the Meridian, will be the Angle of Pofition fought.

Thus, the Angle of Pofition at the *Lizard*, between the Meridian of the *Lizard*

and

and the Great Circle paſſing from thence
to *Barbadoes* is 69 Degrees South-Weſterly;
but the Angle of Poſition between the ſame
Places at *Barbadoes*, is but 38 Degrees North-
Eaſterly.

SCHOLIUM.

The Angle of Poſition between two
Places is a different thing from what is
meant by the Bearings of Places; the *Bearings* of two Places is determined by a ſort
of ſpiral Line called a *Rhumb Line*, paſſing between them in ſuch a manner, as to
make the ſame or equal Angles with all the
Meridians through which it paſſeth : but the
Angle of *Poſition* is the very ſame thing
with what we call the Azimuth in Aſtronomy; both being formed by the Meridian and
a great Circle paſſing thro' the Zenith of a
given Place, and a given Point, either in
the Heavens, then called the Azimuth, or
upon the Earth, then called the Angle of
Poſition.

From hence may be ſhewed the Error of
that Geographical Paradox, *viz.* If a Place
A bears from another B due Weſt, B ſhall
not bear from A due Eaſt. I find this Paradox vindicated by an Author, who at the
ſame time gives us a true Definition of a
Rhumb Line : but his Arguments are un-
geometrical; for if it be admitted that the

Eaſt

Eaft and Weft Lines make the fame Angles
with all the Meridians, through which they
pafs, it will follow that thefe Lines are the
Parallels of Latitude. For the Path defcribed in Travelling from Eaft to Weft, is
the continuation of the Surface of a *Cone*,
whofe Sides are the Radii of the Sphere,
and Bafe the Parallel of Latitude of the
Traveller : and it is evident, that all the Meridians cut the faid Surface at Right (and
therefore at equal) Angles ; whence it
follows, that the Rhumbs of Eaft and Weft
are the Parallels of Latitude ; though the
Cafe may feem different, when we draw inclining Lines (like Meridians) upon Paper,
without carrying our Ideas any further.

PROB. IX. *To find the* Antæci, Periæci, *and*
Antipodes *to any given Place.*

Bring the given Place to the Meridian, and
having found its Latitude, count the fame
Number of Degrees on the Meridian from
the Equator towards the contrary Pole,
and that will give the Place of the *Antæci.*
The Globe being ftill in the fame Pofition,
fet the Hour-Index to XII at Noon, then
turn the Globe about till the Index points to
the lower XII ; the Place which then lies
under the Meridian, having the fame Latitude with the given Place, is the *Periæci*
required. As the Globe now ftands, the
An-

Antipodes of the given Place are under the
fame Point of the Meridian, that its *Antæci* ftood before :   Or, if you reckon 180
Degrees upon the Meridian from the given
Place, that Point will be the *Antipodes.*

Let the given Place be *London* in the Latitude of 51½, Degrees North; that Place
which lies under the fame Meridian, and in
the Latitude of 51½ Degrees South, is the
Antæci : that which lies in the fame Parallel with *London*, and 180 Degrees of Longitude from it, is the *Periæci ;* and the *Antipodes* is that Place whofe Longitude from
London is 180 Degrees, and Latitude 51½
Degrees South.

PROB. X. *The Hour of the Day at one
Place, being given ; to find the corre-
fpondent Hour (or what o'Clock it is at
that Time) in any other Place.*

The Difference of Time betwixt two
Places is the fame with their Difference of
Longitude; wherefore having found their
Difference of Longitude, reduce it into Time,
(by allowing one Hour for every 15 Degrees,
&c.) and if the Place where the Hour is
required lies { Eafterly, } from the Place
{ Wefterly, }
where the Hour is given, { Add
{ Subtract } the Dif-
ference of Longitude reduced into Time

to

{ to
{ from } the Hour given ; and the Sum or
Remainder will accordingly be the Hour required. Or,

Having brought the Place at which the
Hour is given to the Meridian, set the Hour-
Index to the given Hour ; then turn the
Globe about until the Place where the Hour
is required comes to the Meridian, and the
Index will point out the Hour at the said
Place.

Thus when it is *Noon* at *London*, it is

		H.	M.	
	Rome —	0	52	P. M.
At	*Constantinople* —	2	07	P. M.
	Vera-Cruz —	5	30	A. M.
	Pequin in *China* —	7	50	P. M.

P R O B. XI. *The Day of the Month being
given, to find those Places on the Globe
where the Sun will be Vertical, or in the
Zenith, that Day.*

Having found the Sun's Place in the
Ecliptick, bring the same to the Meridian,
and note the Degree over it ; then turning
the Globe round, all Places that pass under
that Degree will have the Sun Vertical that
Day.

H   P R O B.

PROB. XII. *A Place being given in the* Torrid Zone, *to find thofe two Days in which the Sun fhall be Vertical to the fame.*

Bring the given Place to the Meridian, and mark what Degree of Latitude is exactly over it; then turning the Globe about thofe two Points of the Ecliptick, which pafs exactly under the faid Mark, give the Sun's Place at the times required; look upon the Wooden Horizon for thofe two Points of the Ecliptick, and right againft them you'll have the Days required.

PROB. XIII. *To find where the Sun is Vertical at any given Time affigned; or, the Day of the Month and the Hour at any Place (fuppofe* London) *being given, to find in what Place the Sun is Vertical at that very Time.*

Having found the Sun's Declination, and brought the firft Place (*London*) to the Meridian, fet the Index to the given Hour, then turn the Globe about until the Index points to 12 at Noon; which being done, that Place upon the Globe which ftands under the Point of the Sun's Declination upon the Meridian, has the Sun that Moment in the Zenith.

PROB.

PROB. XIV. *The Day, and the Hour of the Day at one Place, being given; to find all thofe Places upon the Earth, where the Sun is then Rifing, Setting, Culminating, (or on the Meridian;) alfo where it is Day-Light, Twilight, Dark Night, Mid-night; where the Twilight then begins, and where it ends: the Height of the Sun in any Part of the illuminated Hemifphere; alfo his Depreffion in the obfcure Hemifphere.*

Having found the Place where the Sun is Vertical at the given Hour, rectify the Globe for the Latitude, and bring the faid Place to the Meridian.

Then all thofe Places that are in the Weftern Semicircle of the Horizon, have the Sun rifing at that Time.

Thofe in the Eaftern Semicircle have it fetting.

To thofe who live under the upper Semicircle of the Meridian, it is 12 o'Clock at Noon. And,

Thofe who live under the lower Semicircle of the Meridian have it Midnight.

All thofe Places that are above the Horizon, have the Sun above them, juft fo much as the Places themfelves are diftant from the Horizon; which Height may be known by fixing the Quadrant of Altitude

in

in the Zenith, and laying it over any particular Place.

In all thofe Places that are 18 Degrees below the Weftern Side of the Horizon, the Twilight is juft beginning in the Morning, or the Day breaks. And in all thofe Places that are 18 Degrees below the Eaftern Side of the Horizon, the Twilight is ending, and the total Darknefs beginning.

The Twilight is in all thofe Places whofe Depreffion below the Horizon does not exceed 18 Degrees. And

All thofe Places that are lower than 18 Degrees have dark Night.

The Depreffion of any Place below the Horizon is equal to the Altitude of its *Antipodes,* which may be eafily found by the Quadrant of Altitude.

PROB. XV. *The Day of the Month being given; to fhew, at one view, the Length of Day and Night in all Places upon the Earth at that Time; and to explain how the Viciffitudes of Day and Night are really made by the Motion of the Earth round her Axis in 24 Hours, the Sun ftanding ftill.*

The Sun always illuminates one half of the Globe, or that Hemifphere which is next towards him, while the other remains in Darknefs: And if (as by the laft Problem) we

we elevate the Globe according to the Sun's
Place in the Ecliptick, it is evident, that the
Sun (he being at an immense Distance from
the Earth) illuminates all that Hemisphere,
which is above the Horizon; the Wooden
Horizon it self will be the Circle terminating Light and Darkness; and all those
Places that are below it, are wholly deprived
of the Solar Light.

The Globe standing in this Position; those
Arches of the Parallels of Latitude which
stand above the Horizon, are the *Diurnal
Arches,* or the Length of the Day in all
those Latitudes at that Time of the Year;
and the remaining Parts of those Parallels,
which are below the Horizon, are the *Nocturnal Arches,* or the Length of the Night
in those Places. The Length of the Diurnal Arches may be found, by counting
how many Hours are contained between the
two Meridians, cutting any Parallel of Latitude, in the Eastern and Western Parts of
the Horizon.

In all those Places that are in the Western
Semicircle of the Horizon, the Sun appears
rising: for, the Sun standing still in the
Vertex, (or above the Brass Meridian) appears Easterly, and 90 Degrees distant from
all those Places that are in the Western Semicircle of the Horizon; and therefore in
those Places he is then rising. Now if
we pitch upon any particular Place upon

the Globe, and bring it to the Meridian, and then bring the Hour-Index to the lower XII, which in this cafe we'll fuppofe to be 12 at Noon, (becaufe otherwife the Numbers upon the Hour-Circle will not anfwer our Purpofe;) and afterwards turn the Globe about, until the aforefaid Place be brought to the Weftern Side of the Horizon; the Index will then fhew the Time of Sun-rifing in that Place. Then turning the Globe gradually about from Weft to Eaft, and minding the Hour-Index, we fhall fee the Progrefs made in the Day every Hour, in all Latitudes upon the Globe, by the real Motion of the Earth round its Axis; until, by their continual Approach to the Brafs Meridian (over which the Sun ftands ftill all the while) they at laft have Noon-Day, and the Sun appears at the higheft; and then by degrees, as they move Eafterly, the Sun feems to decline Weftward, until, as the Places fucceffively arrive in the Eaftern Part of the Horizon, the Sun appears to fet in the Weftern; for the Places that are in the Horizon, are 90 Degrees diftant from the Sun. We may obferve, that all Places upon the Earth, that differ in Latitude, have their Days of different length, (except when the Sun is in the Equinoctial) being longer or fhorter, in proportion to what Part of the Parallels ftand above the Horizon. Thofe that are in the fame Latitude have their

Days

Days of the fame length; but haye them commence fooner or later, according as the Places differ in Longitude.

PROB. XVI. *To explain in general the Al-teration of Seafons, or Length of the Days and Nights, made in all Places of the World, by the Sun's (or the Earth's) annual Motion in the Ecliptick.*

It has been fhewed in the laft Problem, how to place the Globe in fuch a Pofition, as to exhibit the Length of the Diurnal and Nocturnal Arches, in all Places of the Earth, at a partciular Time : If the Hour-Circle be taken off, fo that the Poles of the Globe may be brought to the Horizon; and the Globe be continually rectified, according as the Sun alters his Declination, (which may be known by bringing each Degree of the Ecliptick fucceffively to the Meridian) you'll fee the gradual Increafe or Decreafe made in the Days in all Places of the World, according as a greater or leffer Portion of the Parallels of Latitude ftand above the Horizon. We fhall illuftrate this Problem by Examples taken at different Times of the Year.

1. Let the Sun be in the firft Point of ♋, (which happens on the 10*th* of *June*) that Point being brought to the Meridian, will fhew the Sun's Declination to be 23½ Degrees

H 4　　　　　North;

North; then the Globe muft be rectified to the Latitude of 23½ Degrees; and for the bet-Illuftration of the Problem, let the firft Meridian upon the Globe be brought under the Brafs Meridian. The Globe being in this Pofition, you'll fee at one view the Length of the Days in all Latitudes, by counting the Number of Hours contained between the two extreme Meridians, cutting any particular Parallel you pitch upon, in the Eaftern and Weftern Part of the Horizon. And you may obferve, that the lower Part of the Arctick Circle juft touches the Horizon, and confequently all the People who live in that Latitude have the Sun above their Horizon for the fpace of 24 Hours, without fetting; only when he is in the lower Part of the Meridian (which they would call 12 at Night) he juft touches the Horizon.

To all thofe who live between the Arctick Circle and the Pole, the Sun does not fet, and its Height above the Horizon, when he is in the lower Part of the Meridian, is equal to their Diftance from the Arctick Circle: for Example, Thofe who live in the 80th Parallel have the Sun when he is loweft at this Time 13½ Degrees high.

If we caft our Eye Southward, towards the Equator, we fhall find, that the Diurnal Arches, or the Length of the Days in the feveral Latitudes, gradually leffen: The Diurnal Arch of the Parallel of *London* at

2   this

this Time is 16½ Hours; that of the *Equator*
(is always) 12 Hours; and so continually
less, till we come to the *Antarctick Circle,*
the upper Part of which just touches the
Horizon; and those who live in this Latitude have just one sight of the Sun, peeping
as it were in the Horizon. And all that
Space between the Antarctick Circle and
the South Pole, lies in total Darkness.

If from this Position we gradually move
the Meridian of the Globe, according to the
progressive Alterations made in the Sun's
Declination, by his Motion in the Ecliptick;
we shall find the Diurnal Arches of all those
Parallels, that are on the Northern Side of
the Equator continually decrease; and those
on the Southern Side continually increase,
in the same manner as the Days in those
Places shorten and lengthen. Let us again
observe the Globe when the Sun has got
within 10 Degrees of the Equinoctial; now
the lower Part of the 80*th* Parallel of North
Latitude just touches the Horizon, and all
the Space betwixt this and the Pole, falls in
the illuminated Hemisphere; but all those
Parallels that lie betwixt this and the Arctick
Circle, which before were wholly above
the Horizon, do now intersect it, and the
Sun appears to them to rise and set. From
hence to the Equator, we shall find that
the Days have gradually shortened; and
from the Equator Southward, they have gradually

dually lengthened, until we come to the 80*th* Parallel of South Latitude, the upper part of which juft touches the Horizon, and all Places betwixt this and the South Pole are in total Darknefs: but thofe Parallels betwixt this and the Antarctick Circle, which before were wholly above the Horizon, are now partly above it; the Length of their Days being exactly equal to that of the Nights in the fame Latitude in the contrary Hemifphere. This alfo holds univerfally, that the Length of the Day in one Latitude North, is exactly equal to the Length of the Night in the fame Latitude South; and *vice verfa.*

Let us again follow the Motion of the Sun, until he has got into the Equinoctial, and take a view of the Globe while it is in this Pofition. Now all the Parallels of Latitude are cut into two equal Parts by the Horizon, and confequently the Days and Nights are of equal lengths, *viz.* 12 Hours each in all Places of the World; the Sun rifing and fetting at Six o'Clock, excepting under the two *Poles*, which now lie exactly in the Horizon: Here the Sun feems to ftand ftill in the fame Point of the Heavens for fome Time, until by degrees, by his Motion in the Ecliptick, he afcends higher to one, and difappears to the other, there being properly no Days and Nights under the Poles; for there the Motion of the Earth round its Axis can't be obferved.   If

If we follow the Motion of the Sun towards the Southern Tropick, we fhall fee
the Diurnal Arches of the Northern Parallels
continually decreafe, and the Southern ones
increafe in the fame Proportion, according
to their refpective Latitudes : the North
Pole continually defcending, and the South
Pole afcending above the Horizon, until
the Sun arrives into ♈ at which Time, all
the Space within the Antarctick Circle is
above the Horizon; while the Space between the Arctick Circle and its Neighbouring Pole, is in total Darknefs. And we
fhall now find all other Circumftances quite
reverfe to what they were when the Sun
was in ♋; the Nights now all over the
World, being of the fame length that the
Days were of before.

We have now got to the Extremity of
the Sun's Declination; and if we follow
him thro' the other half of the Ecliptick,
and rectify the Globe accordingly, we fhall
find the Seafons return in their Order, until at length we bring the Globe into its firft
Pofition.

PROB.

PROB. XVII. *To shew by the Globe, at one view, the Length of the Days and Nights in any particular Place, at all Times of the Year.*

Because the Sun by his Motion in the Ecliptick, alters his Declination a small matter every Day; if we suppose all the Torrid Zone to be filled up with a spiral Line, having so many turnings; or a Screw having so many Threads, as the Sun is Days in going from one Tropick to the other; and these Threads at the same Distance from one another in all Places, as the Sun alters his Declination in one Day in all those Places respectively. This Spiral Line or Screw will represent the apparent Paths described by the Sun round the Earth every Day; and by following the Thread from one Tropick to the other, and back again, we shall have the Path the Sun seems to describe round the Earth in a Year. But because the Inclinations of these Threads to one another are but small, we may suppose each Diurnal Path to be one of the Parallels of Latitude, drawn, or supposed to be drawn upon the Globe. Thus much being premised, we shall explain this *Problem*, by placing the Globe according to some of the most remarkable Positions of it; as before we did for the most remarkable Seasons of the Year.

In

In the preceding Problem, the Globe being rectified according to the Sun's Declination, the upper Parts of the Parallels of Latitude, represented the *Diurnal Arches*, or the Length of the *Days* all over the World at that particular Time : Here we are to rectify the Globe according to the Latitude of the Place, and then the upper Parts of the Parallels of Declination are the Diurnal Arches ; and the Length of the Days at all Times of the Year, may be here determined, by finding the Number of Hours contained between the two extreme Meridians, which cut any Parallel of Declination, in the Eastern and Western Points of the Horizon ; after the same manner as before we found the Length of the Day in the several Latitudes at a particular Time of the Year.

1. Let the Place proposed be under the Equinoctial, and let the Globe be accordingly rectified for oo Degrees of Latitude, which is called a direct Position of the Sphere. Here all the Parallels of Latitude, which in this case we'll call the Parallels of Declination, are cut by the Horizon into two equal Parts ; and consequently those who live under the Equinoctial have the Days and Nights of the same Length at all Times of the Year ; also in this Part of the Earth, all the *Stars* rise and set, and their Continuance above the Horizon, is equal to their Stay below it, *viz.* 12 Hours.

If

If from this Position we gradually move the Globe according to the several Alterations of Latitudes, which we will suppose to be Northerly; the Lengths of the Diurnal Arches will continually increase, until we come to a Parallel of Declination, as far distant from the Equinoctial, as the Place it self is from the Pole. This Parallel will juft touch the Horizon, and all the Heavenly Bodies that are betwixt it and the Pole never descend below the Horizon. In the mean Time, while we are moving the Globe, the Lengths of the Diurnal Arches of the *Southern* Parallels of Declination, continually diminish in the same Proportion that the Northern ones increafed; until we come to that Parallel of Declination which is so far diftant from the Equinoctial Southerly, as the Place it self is from the North Pole. The upper part of this *Parallel* juft touches the Horizon, and all the Stars that are betwixt it and the South Pole, never appear above the Horizon. All the Nocturnal Arches of the Southern Parallels of Declination, are exactly of the same Length with the Diurnal Arches of the correspondent Parallels of North Declination.

2. Let us take a view of the Globe, when it is rectified for the Latitude of *London*, or 51½ Degrees North. When the Sun is in the Tropick of ♋, the Day is about 16¼ Hours; as he recedes from this Tropick,

the

the Days proportionably shorten, until he arrives into ♑, and then the Days are at the shortest, being now of the same length with the Night when the Sun was in ♋, *viz.* 7¼ Hours. The lower part of that Parallel of Declination, which is 38½ Degrees from the Equinoctial Northerly, just touches the Horizon ; and all the Stars that are betwixt this Parallel and the North Pole, never set to us at *London.* In like manner the upper part of the Southern Parallel of 38½ Degrees just touches the Horizon, and all the Stars that lie betwixt this Parallel and the South Pole are never visible in his Latitude.

Again, let us rectify the Globe for the Latitude of the *Arctick Circle,* we shall then find, that when the Sun is in ♋, he touches the Horizon on that Day, without setting, being 24 Hours compleat above the Horizon; and when he is in *Capricorn,* he once appears in the Horizon, but does not rise for the space of 24 Hours : when he is in any other Point of the Ecliptick, the Days are longer or shorter, according to his Distance from the Tropicks. All the Stars that lie between the Tropick of *Cancer,* and the North Pole, never set in this Latitude ; and those that are between the Tropick of *Capricorn* and the South Pole, are always hid below the Horizon.

If we elevate the Globe still higher, the Circle of *perpetual Apparition* will be nearer
the

the Equator, as will that of *perpetual Occultation* on the other side. For Example, Let us rectify the Globe for the Latitude of 80 Degrees North; when the Sun's Declination is 10 Degrees North, he begins to turn above the Horizon without setting, and all the while he is making his Progress from this Point to the Tropick of ♋, and back again, he never sets. After the same manner, when his Declination is 10 Degrees South, he is just seen at Noon in the Horizon; and all the while he is going Southward, and back again, he disappears, being hid just so long as before, at the opposite Time of the Year he appeared visible.

Let us now bring the North Pole into the Zenith, then will the Equinoctial coincide with the Horizon; and consequently all the Northern Parallels are above the Horizon, and the Southern ones below it. Here is but one Day and one Night throughout the Year; it being Day all the while the Sun is to the Northward of the Equinoctial, and Night for the other half Year. All the Stars that have North Declination always appear above the Horizon, and at the same Height; and all those that are on the other side, are never seen.

What has been here said of rectifying the Globe to North Latitude, holds for the same Latitude South; only that before the longest Days were, when the Sun was in ♋, the
same

same happening now when the Sun is in ♈;
and so of the rest of the Parallels, the Seasons
being directly opposite to those who live in
different Hemispheres.

The three foregoing Problems are very
useful, as they give us a general Idea how
the Seasons are altered in all Places of the
World, and do very well deserve the particular Attention of the Reader; it being
undoubtedly entertaining, to know how
these things are brought about by the re-
gular Course of Nature.

For the sake of the Reader, I shall explain
some Things delivered above in general
Terms, by particular Problems.

But from what has been already said, we
may first make the following Observations.

1. *All Places of the Earth do equally
enjoy the benefit of the Sun, in respect of
Time, and are equally deprived of it; the
Days at one Time of the Year being exactly
equal to the Nights at the opposite Season.*

2. *In all Places of the Earth, save exactly under the Poles, the Days and Nights
are of equal Length, (viz. 12 Hours each)
when the Sun is in the Equinoctial.*

3. *Those who live under the Equinoctial,
have the Days and Nights of equal lengths,
at all Times of the Year.*

4. *In all Places between the Equinoctial
and the Poles, the Days and Nights are*

I   *never*

never equal, but when the Sun is in the Equinoctial Points ♈ and ♎.

5. The nearer any Place is to the Equator, the less is the Difference between the Length of the Artificial Days and Nights in the said Place; and the more remote, the greater.

6. To all the Inhabitants. lying under the same Parallel of Latitude, the Days and Nights are of equal Lengths, and that at all Times of the Year.

7. The Sun is Vertical twice a Year, to all Places between the Tropicks; to those under the Tropicks, once a Year, but never any where else.

8. In all Places between the Polar Circles, and the Poles, the Sun appears some Number of Days without setting; and at the opposite Time of the Year, he is for the same length of Time without rising: and the nearer unto, or further remote from the Pole, those Places are, the longer or shorter is the Sun's continued Presence in, or Absence from the same.

9. In all Places lying exactly under the Polar Circles, the Sun, when he is in the nearest Tropick, appears 24 Hours without setting; and when he is in the contrary Tropick, he is for the same length of Time without rising; but at all other Times of the Year he rises and sets there, as in other Places.

10.

10. *In all Places lying in the* $\left\{\begin{array}{l}Northern\\Southern\end{array}\right\}$
Hemisphere, the Longest Day, and Shortest Night, is when the Sun is in the $\left\{\begin{array}{l}Northern\\Southern\end{array}\right\}$ *Tropick; and the contrary.*

PROB. XVIII. *The Latitude of any Place, not exceeding* 66½ *Degrees, and the Day of the Month being given, to find the Time of Sun-rising and setting, and the Length of the Day and Night.*

Having rectified the Globe according to the Latitude, bring the Sun's Place to the Meridian, and put the Hour-Index to 12 at Noon; then bring the Sun's Place to the Eastern Part of the Horizon, and the Index will shew the Time when the Sun rises. Again, turn the Globe until the Sun's Place be brought to the Western Side of the Horizon, and the Index will shew the Time of Sun-setting.

The Hour of Sun-setting doubled, gives the Length of the Day; and the Hour of Sun-rising doubled, gives the Length of the Night.

Let it be required to find when the Sun rises and sets at *London* on the 20*th* of *April.* Rectify the Globe for the Latitude of *London,* and having found the Sun's Place, corresponding to *April* the 20*th, viz*

I 2   ♂

♉ 10¾ Degree, bring ♉ 10¾ Degrees to the
Meridian, and ſet the Index to 12 at Noon ;
then turn the Globe about till ♉ 10¾ Degrees
be brought to the Eaſtern part of the Horizon, and you'll find the Index point 4¾ Hours ;
this being doubled, gives the Length of the
Night 9½ Hours.   Again, bring the Sun's
Place to the Weſtern part of the Horizon,
and the Index will point 7¼ Hours, which
is the Time of Sun-ſetting ; this being
doubled, gives the Length of the Day 14½
Hours.

PROB. XIX. *To find the Length of the
Longeſt and Shorteſt Day and Night
in any given Place, not exceeding 66½
Degrees of Latitude.*

Note, The Longeſt Day at all Places on
the {North / South} Side of the Equator, is when

the Sun is in the firſt Point of {Cancer / Capricorn :}
Wherefore having rectified the Globe for the
Latitude, find the Time of Sun-riſing and
ſetting, and thence the Length of the Day
and Night as in the laſt Problem, according
to the Place of the Sun : Or having rectified
the Globe for the Latitude, bring the ſolſti-
tial Point of that Hemiſphere to the Eaſt
part of the Horizon, and ſet the Index to 12
at Noon; then turning the Globe about till

2   the

the said Solstitial Point touches the Western Side of the Horizon ; the Number of Hours from Noon to the Place where the Index points (being counted according to the Motion of the Index) is the Length of the Longest Day ; the Complement whereof to 24 Hours, is the Length of the Shortest Night, and the Reverse gives the Shortest Day and the longest Night.

	Longest Day.	*Short.N.*
	Deg. Hours.	Hours.
Thus in Lat.	45 —15$\frac{1}{2}$	8$\frac{1}{2}$
	51$\frac{1}{2}$—16$\frac{1}{2}$	7$\frac{1}{2}$
	60 —18$\frac{1}{2}$	5$\frac{1}{2}$

If from the Length of the Longest Day, you subtract 12 Hours, the Number of Half-Hours remaining will be the *Climate* : Thus, that Place where the longest Day is 16$\frac{1}{2}$Hours, lies in the 9th *Climate*. And by the Reverse, having the *Climate*, you have thereby the Length of the Longest Day.

PROB. XX, *To find in what Latitude the Longest Day is, of any given Length less than 24 Hours.*

Bring the Solstitial Point to the Meridian, and set the Index to 12 at Noon ; then turn the Globe Westward till the Index points at half the Number of Hours given : which being done, keep the Globe from turning round its Axis, and slide the Meri-

dian

dian up or down in the Notches, till the
Solftitial Point comes to the Horizon, then
that Elevation of the Pole will be the Latitude.

If the Hours given be 16, the Latitude
is 49 Degrees; if 20 Hours, the Latitude is
$63\frac{1}{4}$ Degrees.

P R O B. XXI. *A Place being given in one
of the* Frigid Zones *(suppose the* Northern) *to find what Number of Days
(of 24 Hours each) the Sun doth con-
stantly shine upon the same, how long he
is absent, and also the first and last Day
of his Appearance.*

Having rectified the Globe according to
the Latitude, turn it about until some Point
in the first Quadrant of the Ecliptick (because the Latitude is North) interfect the
Meridian in the North Point of the Horizon;
and right against that Point of the Ecliptick
on the Horizon, stands the Day of the
Month when the Longest Day begins.

And if the Globe be turned about till
some Point in the second Quadrant of the
Ecliptick cuts the Meridian in the same Point
of the Horizon, it will shew the Sun's Place
when the longest Day ends; whence the
Day of the Month may be found as before.
Then the Number of Natural Days contain'd between the times the longest Day

begins

begins and ends, is the Length of the Longeft Day required.

Again turn the Globe about, until fome
Point in the third Quadrant of the Ecliptick
cuts the Meridian in the South Part of the
Horizon; that Point of the Ecliptick will
give the Time when the Longeft Night begins. Laftly, turn the Globe about until
fome Point in the fourth Quadrant of the
Ecliptick cuts the Meridian in the South
Point of the Horizon; and that Point of the
Ecliptick will be the Place of the Sun, when
the Longeft Night ends.

Or, the Time when the Longeft Day or
Night begins, being known, their end may
be found by counting the Number of Days
from that Time to the fucceeding Solftice;
then counting the fame Number of Days
from the Solftitial Day, will give the Time
when it ends.

Or, if you bring the Solftitial Point ♋ to
the North Part of the Meridian, then keep
the Globe in that Pofition; the Place where
the Ecliptick cuts the North-Eaftern and
North-Weftern Parts of the Horizon, will be
the Sun's Place when the Longeft Day begins and ends; and where it cuts the South-
Weftern, and. South-Eaftern Parts of the
Horizon, will be the Place of the Sun when
the longeft Night begins and ends.

Thus at the *North Cape* on the Coaft of
Lapland, in the Latitude of 71½ Degrees,

the longeſt Day begins about the 4th of
May, and ends the 19th of *July*; after
which, the Sun riſes and ſets till the 4th of
November, when he but firſt touches the
Horizon in the Southermoſt Point of it, and
then continues below it till the 16th of *Ja-
nuary*, when he'll juſt appear to riſe in the
Meridian after he had been hid below the
Horizon, for the Space of 73 *Natural* Days,
which is the Length of the Longeſt Night.

After the ſame manner may theſe things
be eaſily found for any Place either within
the *Arctick* or *Antarctick Circle*.

PROB. XXII. *To find in what Latitude
the Longeſt Day is of any given Length
leſs than* 182 *Natural Days.*

Find a Point in the Ecliptick, half ſo
many Degrees diſtant from the Solſtitial
Point, as there are Days given, and bring
that Point to the Meridian; then keep the
Globe from turning round its Axis, and
move the Meridian up or down until
the foreſaid Point of the Ecliptick comes
to the Horizon: that Elevation of the Pole
will be the Latitude required.

If the Days given were 78, the Latitude
is 71½ Degrees.

This Method is not accurate, becauſe the
Degrees in the Ecliptick do not correſpond
to Natural Days; and alſo becauſe the
Sun

Sun does not always move in the Ecliptick
at the fame rate; however fuch Problems
as thefe may ferve for Amufements.

PROB. XXIII. *The Day of the Month being
given, to find when the Morning and
Evening* Twilight *begins and ends, in
any Place upon the Globe.*

In the aforegoing Problems, by the Length
of the Day, we meant the Time from Sun-
rifing to Sun-fet; and the Night we reckoned
from Sun-fet till he rofe next Morning. But
it is found by experience, that *Total Darknefs* does not commence in the Evening,
till the Sun has got 18 Degrees below the
Horizon; and when he comes within the
fame Diftance of the Horizon next Morning,
we have the firft *Dawn of Day.* This
faint Light which we have in the Morning
and Evening before and after the Sun's rifing
and fetting, is what we call the *Twilight.*
　*Having rectified the Globe for the Lati- *Prob.VI.
tude, the Zenith, and the Sun's Place; turn
the Globe, and the Quadrant of Altitude
until the Sun's Place cuts 18 Degrees below
the Horizon (if the Quadrant reaches fo far)
then theIndex upon the Hour-Circle willfhew
the Beginning or Ending of Twilight, after
the fame manner as before we found the
Time of Sun-rifing and fetting, in *Prob.* 18.
But by reafon of the Thicknefs of theWooden
Hori-

Horizon, we can't conveniently fee, or compute when the Sun's Place is brought to the Point aforefaid. Wherefore the Globe being rectified as above directed, turn the Globe, and alfo the Quadrant of Altitude Westward, until that Point in the Ecliptick, which is oppofite to the Sun's Place, cuts the Quadrant in the 18*th* Degree above the Horizon; then the Hour-Index will fhew the Time when Day breaks in the Morning. And if you turn the Globe and the Quadrant of Altitude, until the Point oppofite to the Sun's Place cuts the Quadrant in the 18*th* Degree in the Eaftern Hemifphere; the Hour-Hand will fhew when Twilight ends in the Evening. Or, having found the Time from Midnight when the Morning Twilight begins, if you reckon fo many Hours before Midnight, it will give the Time when the Evening Twilight ends. Having found the Time when Twilight begins in the Morning, find the Time of Sun-rifing, by *Prob.* 18. and the Difference will be the Duration of Twilight.

Thus at *London*, on the firft of *May*, Twilight begins at three quarters paft One o'Clock; the Sun rifes at about half an Hour paft Four; whence the Duration of Twilight now is 2¾ Hours, both in the Morning and Evening. On the firft of *November*, the Twilight begins at half an Hour paft Six, being fomewhat above an Hour before Sun-rifing.

PROB.

PROB. XXIV. *To find the Time when total Darkness ceases, or when the Twilight continues from Sun-setting to Sun-rising, in any given Place.*

Let the Place be in the Northern Hemisphere; then if the Complement of the Latitude be greater than (the Depression) 18 Degrees, subtract 18 Degrees from it, and the Remainder will be the Sun's Declination North, when total Darkness ceases. But if the Compliment of the Latitude is less than 18 Degrees, their Difference will be the Sun's Declination South, when the Twilight begins to continue all Night. If the Latitude is South, the only Difference will be, that the Sun's Declination will be on the contrary side.

Thus at *London,* when the Sun's Declination North is greater than 20¼ Degrees, there is no total Darkness, but constant Twilight, which happens from the 15*th* of *May* to the 7*th* of *July,* being near two Months. Under the North Pole the Twilight ceases, when the Sun's Declination is greater than 18 Degrees South; which is from the 2*d* of *November,* till the 18*th* of *January:* so that notwithstanding the Sun is absent in this part of the World for half a Year together, yet total Darkness does not continue above 11 Weeks; and besides, the

Moon

Moon is above the Horizon once a Month for a whole Fortnight, throughout the Year.

PROB. XXV. *The Day of the Month being given, to find thofe Places of the Frigid Zones, where the Sun begins to fhine conftantly without fetting; and alfo thofe Places where he begins to be totally abfent.*

Bring the Sun's Place to the Meridian, and mark the Number of Degrees contained betwixt that Point and the Equator; then count the fame Number of Degrees from the neareft Pole (*viz.* the North Pole, if the Sun's Declination is Northerly, other-wife the South Pole) towards the Equator, and note that Point upon the Meridian; then turn the Globe about, and all the Places which pafs under the faid Point, are thofe where the Sun begins to fhine con-ftantly, without fetting on the given Day. If you lay the fame Diftance from the op-pofite Pole towards the Equator, and turn the Globe about, all the Places which pafs under that Point, will be thofe where the longeft Night begins.

PROB. XXVI. *The Latitude of the Place being given, to find the Hour of the Day when the Sun fhines.*

If it be in the Summer, elevate the Pole according to the Latitude, and fet the Meridian

ridian due North and South ; then the Shadow of the Axis will cut the Hour on the
Dial-Plate : For the Globe being rectified
in this manner, the Hour-Circle is a true
Equinoctial Dial; the Axis of the Globe
being the *Gnomon.* This holds true in
Theory, but it might not be very accurate
in Practice, because of the Difficulty in
placing the Horizon of the Globe truly Horizontal, and its Meridian due North and
South.

If it be in the Winter Half-Year, elevate
the South Pole according to the Latitude
North ; and let the North Part of the Horizon be in the South Part of the Meridian :
then the Shade of the Axis will shew the
Hour of the Day as before. But this cannot be so conveniently performed, tho' the
Reason is the same as in the former Case.

P R O B. XXVII. *To find the Sun's Altitude
when it shines, by the Globe.*

Having set the Frame of the Globe truly
horizontal or level, turn the North Pole
towards the Sun, and move the Meridian
up or down in the Notches, till the Axis
casts no Shadow; then the Arch of the
Meridian, contained betwixt the Pole and
the Horizon, is the Sun's Altitude.

Note, The best way to find the Sun's
Altitude, is by a little Quadrant graduated
into

into Degrees, and having Sights and a Plummet to it: Thus, hold the Quadrant in your Hand, fo as the Rays of the Sun may pafs thro' both the Sights ; the Plummet then hanging freely by the Side of the Inftrument, will cut in the Limb the Altitude required. Thefe Quadrants are to be had at the Inftrument-Makers, with Lines drawn upon them, for finding the Hour of the Day, and the Azimuth, with feveral other pretty Conclufions, very entertaining for Beginners.

PROB. XXVIII. *The Latitude, and the Day of the Month being given, to find the Hour of the Day when the Sun fhines.*

Having placed the Wooden Frame upon a Level, and the Meridian due North and South, rectify the Globe for the Latitude, and fix a Needle perpendicularly over the Sun's Place: The Sun s Place being brought to the Meridian, fet the Hour-Index to 12 at Noon, then turn the Globe about until the Needle points exactly to the Sun, and cafts no Shadow, and then the Index will fhew the Hour of the Day.

PROB.

PROB. XXIX. *The Latitude, the Sun's Place, and his Altitude, being given; to find the Hour of the Day, and the Sun's Azimuth from the Meridian.*

Having rectified the Globe for the Latitude, the Zenith, and the Sun's Place; turn the Globe, and the Quadrant of Altitude, so that the Sun's Place may cut the given Degree of Altitude; then the Index will shew the Hour, and the Quadrant will cut the Azimuth in the Horizon. Thus, If at *London*, on the 10*th* of *August*, the Sun's Altitude be 36 Degrees in the Forenoon, the Hour of the Day will be IX, and the Sun's Azimuth about 58 Degrees from the South Part of the Meridian.

PROB. XXX. *The Sun's Azimuth being given, to place the Meridian of the Globe due North and South, or to find a Meridian Line when the Sun shines.*

Let the Sun's Azimuth be 30 Degrees South-Easterly, set the Horizon of the Globe upon a level, and bring the North Pole into the Zenith; then turn the Horizon about, until the Shade of the Axis cuts as many Hours as is equivalent to the Azimuth, (allowing 15 Degrees to an Hour) in the North-West Part of the Hour-Circle;

2   *viz.*

viz. X at Night; which being done, the Meridian of the Globe ſtands in the true Meridian of the Place. The Globe ſtanding in this Poſition, if you hang two Plummets at the North and South Points of the Wooden Horizon, and draw a Line betwixt them, you'll have a Meridian Line; which if it be on a fixed Plain (as a Floor or Window) it will be a Guide for placing the Globe due North and South at any other Time.

PROB. XXXI. *The Latitude, Hour of the Day, and the Sun's Place being given, to find the Sun's Altitude and Azimuth.*

Rectify the Globe for the Latitude, the Zenith, and the Sun's Place; then the Number of Degrees contained betwixt the Sun's Place and the Vertex is the Sun's Meridional Zenith Diſtance; the Complement of which, to 90 Degrees, is the Sun's Meridian Altitude. If you turn the Globe about until the Index points at any other given Hour, then bringing the Quadrant of Altitude to cut the Sun's Place, you'll have the Sun's Altitude at that Hour; and where the Quadrant cuts the Horizon, is the Sun's Azimuth at the ſame Time. Thus *May* the 20*th*, at *London*, the Sun's Meridian Altitude will be 61¼ Degrees; and at 10 o'Clock in the Morning, the Sun's Altitude will be 52 Degrees; and his Azimuth about 50 Degrees from the South Part of the Meridian.

PROB.

PROB. XXXII. *The Latitude of the Place,*
and the Day of the Month being given;
to find the Depreſſion of the Sun below
the Horizon, and his Azimuth at any
Hour of the Night.

Having rectified the Globe for the Latitude,
the Zenith, and the Sun's Place; take a Point
in the Ecliptick, exactly oppoſite to the Sun's
Place, and find the Sun's Altitude, and Azimuth, as by the laſt Problem; and theſe
will be the Depreſſion and the Altitude required. Thus, If the Time given be the
20*th* of *November*, at 10 o'Clock at Night,
the Depreſſion and Azimuth will be the
ſame as was found in the laſt Problem.

PROB. XXXIII. *The Latitude, the Sun's*
Place, and his Azimuth being given;
to find his Altitude, and the Hour.

Rectify the Globe for the Latitude, the
Zenith, and Sun's Place; then put the
Quadrant of Altitude to the Sun's Azimuth
in the Horizon, and turn the Globe till the
Sun's Place meet the Edge of the Quadrant;
then the ſaid Edge will ſhew the Altitude,
and the Index point to the Hour. Thus,
May the 10*th* at *London*, when the Sun is due
Eaſt, his Altitude will be about 24 Degrees,
and the Hour about VII in the Morning:

and when his Azimuth is 60 Degrees South-Wefterly, the Altitude will be about 44½ Degrees, and the Hour about 2¾ in the Afternoon.

Thus, the Latitude and the Day being known, and having befides either the Altitude, the Azimuth, or the Hour; the other two may be eafily found.

PROB. XXXIV. *The Latitude, the Sun's Altitude, and his Azimuth, being given; to find his Place in the Ecliptick and the Hour.*

Rectify the Globe for the Latitude and Zenith, and fet the Edge of the Quadrant, to the given Azimuth; then turning the Globe about, that Point of the Ecliptick which cuts the Altitude, will be the Sun's Place. Keep the Quadrant of Altitude in the fame Pofition, and having brought the Sun's Place to the Meridian, and the Hour-Index to 12 at Noon; turn the Globe about till the Sun's Place cuts the Quadrant of Altitude, and then the Index will point the Hour of the Day.

PROB. XXXV. *The Declination and Meridian Altitude of the Sun, or of any Star being given; to find the Latitude of the Place.*

Mark the Point of Declination upon the Meridian, according as it is either North or South,

South, from the Equator; then flide the Meridian up or down in the Notches, till the Point of Declination be fo far diftant from the Horizon, as is the given Meridian Altitude; that Elevation of the Pole will be the Latitude.

Thus, If the Sun's, or any Star's Meridian Altitude be 50 Degrees South, and its Declination 11½ Degrees North, the Latitude will be 51½ Degrees North.

P R O B. XXXVI. *The Day and Hour of a Lunar Eclipfe being known; to find all thofe Plaçes upon the Globe in which the fame will be vifible.*

* Find where the Sun is Vertical at the *Prob.13.* given Hour, and bring that Point to the Zenith; then the Eclipfe will be vifible in all thofe Places that are under the Horizon. Or, if you bring the Antipodes to the Place where the Sun is Vertical, into the Zenith, you'll have the Places where the Eclipfe will be vifible above the Horizon.

Note, Becaufe *Lunar* Eclipfes continue fometimes for a long while together, they may be feen in more Places than one Hemifphere of the Earth; for by the Earth's Motion round its Axis, during the Time of the Eclipfe, the *Moon* will rife in feveral Places after the Eclipfe began.

Note, When an Eclipfe of the *Sun* is Central, if you bring the Place where the

Sun is Vertical at that Time, into the Zenith, some part of the Eclipse will be visible in most Places within the upper Hemisphere : but by reason of the short Duration
of Solar Eclipses, and the Latitude which
the Moon commonly has at that Time,
(tho' but small) there is no certainty in
determining the Places where those Eclipses
will be visible, by the Globe ; but recourse
must be had to Calculations.

PROB. XXXVII. *The Day of the Month,
and Hour of the Day, according to our
way of reckoning in* England, *being given,
to find thereby the* Babylonick, Italick,
and the Jewish *or* Judaical *Hour.*

1. To find the *Babylonick Hour* (which
is the Number of Hours from Sun-rising.)
Having found the Time of Sun-rising in
the given Place, the Difference betwixt this
and the Hour given, is the *Babylonick* Hour.

2. To find the *Italick Hour,* (which is
the Number of Hours from Sun-setting.)
Subtract the Hour of Sun-setting, from the
given Hour, and the Remainder will be the
Italick Hour required.

3. To find the *Jewish Hour,* (which is
$\frac{1}{12}$ Part of an *Artificial* Day.) Find how
many Hours the Day consists of; then say,
as the Number of Hours the Day consists
of, is to 12 Hours ; so is the Hour since Sun-
rising, to the *Judaical* Hour required.

Thus,

Thus, If the Sun rifes at 4 o'Clock, (confequently fets at 8) and the Hour given be 5 in the Evening, the *Babylonick* Hour will be the 13*th* ; the *Italick* the 21*ſt* ; and the *Jewiſh* Hour will be Nine and three Quarters.

The Converſe being given, the Hour of the Day, according to our way of reckoning in *England* may be eafily found.

The following Problems are peculiar to the *Celeſtial* Globe.

PROB. XXXVIII. *To find the Right Aſcenſion and Declination of the Sun or any Fixed Star.*

Bring the Sun's Place in the Ecliptick to the Meridian, then that Degree of the E-quator, which is cut by the Meridian, will be the *Sun's Right Aſcenſion* ; and that De-gree of the Meridian, which is exactly over the Sun's Place, is the *Sun's Declination.* After the fame manner, bring the Place of any Fixed Star to the Meridian, and you'll find its Right Aſcenſion in the Equinoctial, and Declination on the Meridian.

Thus, the Right Aſcenſion and Declina-tion is found, after the fame manner as the Longitude and Latitude of a Place upon the *Terreſtrial* Globe.

Note, The Right Aſcenſion and Decli-nation of the Sun vary every Day; but the

K 3   Right

Right Afcenſion, &c. of the Fixed Stars
is the ſame throughout the Year. *

The Sun's Right Afcenſion. Declin.

		Deg.	*Deg.*
January 20	——	314 —-	17⅓ S.
March 25	——	14 —	6 N.
July 10	——	120¼ ·	20¼ N.
Novemb. 15	——	242 —	21 S.

Thus on

	R. Aſc.	Decl.
	Deg.	*Deg.*
Aldebaran ——	65	16 N.
Spica Virginis —	197¾	9¾ S.
Capella —— ——	74	45⅔ N.
Syrius, or the Dog-Star	98¼	16⅓ S.

Note, The Declination of the Sun may
be found after the ſame manner, by the
Terreſtrial Globe; and alſo his Right Afcenſion, when the Equinoctial is numbered
into 360 Degrees, commencing at the Equinoctial Point ♈ : but as the Equinoctial is
not always numbered ſo; and this being
properly a Problem in *Aſtronomy*, we chuſe
rather to place it here.

By the Converſe of this Problem, having
the Right Afcenſion and Declination of any

---

* The inſenſible Change in the Longitude, Right Afcenſion,
and Declination of the Fixed Stars, made by their ſlow Motion, parallel to the Ecliptick (being but 1 Degree in 72 Years)
is not worth notice in this Place.

Point given, that Point it felf may be eafily found upon the Globe.

P R O B. XXXIX. *To find the Longitude and Latitude of a given Star.*

Having brought the Solſtitial Colure to the Meridian, fix the Quadrant of Altitude over the proper Pole of the Ecliptick, whether it be North or South; then turn the Quadrant over the given Star; and the Arch contained betwixt the Star and the Ecliptick, will be the Latitude, and the Degree cut on the Ecliptick will be the Star's Longitude.

Thus, the Latitude of *Arcturus* will be found to be 31 Degrees North, and the Longitude 200 Degrees from Υ, or 20 Degrees from $\simeq$: The Latitude of *Fomalhaut* in the Southern Fiſh, 21 Degrees South, and Longitude $299\frac{1}{2}$ Degrees, or $\approx 29\frac{1}{2}$ Degrees. By the Converfe of this Method, having the Latitude and Longitude of a Star given, it will be eafy to find the Star upon the Globe.

The Diſtance betwixt two Stars, or the Number of Degrees contained between them, may be found, by laying the Quadrant of Altitude over each of them, and counting the Number of Degrees intercepted; after the fame manner, as we found the Diſtance betwixt two Places on the *Terreſtrial* Globe, in *Prob.* VII.

K 4   P R O B.

PROB. XL. *The Latitude of the Place,*
the Day of the Month, and the Hour
being given; to find what Stars are
then rising or setting, what Stars are Cul-
minating or on the Meridian, and the Al-
titude and Azimuth of any Star above
the Horizon: and also how to distin-
guish the Stars in the Heavens one from
the other, and to know them by their
proper Names.

Having rectified the Globe for the Latitude, the Zenith, and the Sun's Place; turn
the Globe about until the Index points to
the given Hour: the Globe being kept in
this Position.

All those Stars that are in the $\begin{Bmatrix} \text{Eastern} \\ \text{Western} \end{Bmatrix}$

Side of the Horizon, are then $\begin{Bmatrix} \text{Rising} \\ \text{Setting.} \end{Bmatrix}$

All those Stars that are under the Meridian
are then Culminating. And if the Quadrant of Altitude be laid over the Center of
any particular Star, it will shew that Star's
Altitude at that Time, and where it cuts the
Horizon will be the Star's Azimuth from the
North or South part of the Meridian.

The Globe being kept in the same Elevation, and from turning round its Axis;
move the wooden Frame about until the
North and South Points of the Horizon lie
exactly

exactly in the Meridian ; then right Lines
imagined to pafs from the Centre thro' each
Star upon the Surface of the Globe, will
point out the real Stars in the Heavens, which
thofe on the Globe are made to reprefent.
And if you are by the fide of fome Wall,
whofe Bearing you know, lay the Quadrant
of Altitude to that Bearing in the Horizon, and it will cut all thofe Stars, which at
that very Time are to be feen in the fame
Direction, or clofe by the fide of the faid
Wall.   Thus knowing fome of the remarkable Stars in any part of the Heavens, the
Neighbouring Stars may be diftinguifhed, by
obferving their Situations with refpect to
thofe that are already known, and comparing them with the Stars drawn upon the
Globe.

Thus, if you turn your Face towards the
North, you will find the North Pole of the
Globe points to the *Pole-Star* ; then you
may obferve two Stars fomewhat lefs bright
than the Pole-Star, almoft in a right Line
with it, and four more which form a fort of
a *Quadrangle* ; thefe Seven Stars make the
Conftellation called the *Little Bear*, the
Pole-Star being in the Tip of the Tail.   In
this Neighbourhood you'll obferve feven
bright Stars which are commonly called
Charles's Wane ; thefe are the bright Stars
in the *Great Bear*, and do form much fuch
another Figure with thofe before mentioned

in

in the Little Bear : the two foremoſt of the
Square lie almoſt in a right Line with the
Pole-Star, and are called the *Pointers* ; ſo
that knowing the Pointers, you may eaſily
find the Pole-Star. Thus the reſt of the
Stars in this Conſtellation, and all the Stars
in the Neighbouring Conſtellations, may be
eaſily found, by obſerving how the unknown
Stars lie either in *Quadrangles*, *Triangles*
or Streight Lines, with thoſe that are already known upon the Globe.

After the ſame manner the Globe being
rectified, you may diſtinguiſh thoſe Stars
that are to the Southward of you, and be
ſoon acquainted with all the Stars that are
viſible in our Hemiſphere.

SCHOLIUM.

The Globe being rectified to the Latitude
of any Place, if you turn it round its Axis ;
all thoſe Stars that do not go below the Horizon during a whole Revolution of the
Globe, never ſet in that Place ; and thoſe
that do not come above the Horizon, never
riſe.

P R O B. XLI. *The Latitude of the Place
being given; to find the Amplitude,
Oblique Aſcenſion and Deſcenſion, A-
ſcenſional Difference ; Semi-Diurnal
Arch, and the Time of Continuance a-
bove*

*bove the Horizon, of any given Point in
the Heavens.*

Having rectified the Globe for the Latitude, and brought the given *Point* to the
Meridian, set the Index to the Hour of 12 ;
then turn the Globe until the given Point
be brought to the Eastern side of the Horizon, and that Degree of the Equinoctial
which is cut by the Horizon at that Time,
will be the *Oblique Afcenfion*; and where
the given Point cuts the Horizon, is the
Amplitude Ortive : If the Globe be turned
about until the given Point be brought to
the Western Side of the Horizon, it will
there shew the *Amplitude Occafive ;* and
where the Horizon cuts the Equinoctial at
that Time, is the *Oblique Defcenfion.*

The Time between the Index at either of
these two Positions, and the Hour of 6, or
the Difference between the Oblique Afcenfion and Defcenfion, is the *Afcenfional
Difference.*

If the Place be in North Latitude, and the
Declination of the given Point be $\left\{ \begin{matrix} North \\ South \end{matrix} \right\}$
the Afcenfional Difference reduced into
Time, and $\left\{ \begin{matrix} added\ to \\ fubftracted\ from \end{matrix} \right\}$ 6 o' Clock,
gives the *Semi-Diurnal Arch ;* the Comple-
ment whereof to a Semicircle, is the *Se-
mi-Nocturnal Arch.* If the Place be in
South

South Latitude, then the contrary is to be observed with respect to the Declination.

The Semi-${\begin{cases}\text{Diurnal}\\\text{Nocturnal}\end{cases}}$ Arch being doubled, gives the Time of Continuance ${\begin{cases}\text{above}\\\text{below}\end{cases}}$ the Horizon. Or the Time of Continuance above the Horizon, may be found by counting the Number of Hours contained in the upper Part of the Horary Circle, betwixt the Places where the Index pointed when the given Point was in the Eastern and Western Parts of the Horizon. If the given Point was the Sun's Place, the Index pointed the Time of his Rising and Setting, when the said Place was in the Eastern and Western Parts of the Horizon, as in *Prob.* 18. Or the Time of Sun-rising may be found by adding or subtracting his Ascensional Difference, to or from the Hour of Six, according as the Latitude and Declination are either contrary or the same way.

Thus, at *London* on the 20th of *April,* the *Sun's*

Amplitude, is 24 Degrees Northerly.

Oblique Ascension, 20.

Oblique Descension, 58.

Ascensional Difference, 19.

Semidiurnal Arch, 109.

His Continuance above the Horizon 141 Hours.

Sun

Sun rifes at three Quarters paft Four.
Sun fets at a Quarter paft Seven.

Thefe things for the Sun vary every Day ; but for a fixed Star, the Day of the Month need not be given, for they are the fame all the Year round.

In the Latitude of 51½ North, *Syrius's Amplitude*, is about 28 Degrees Southerly.
Oblique Afcenfion, 121.
Oblique Defcenfion, 75.
Afcenfional Difference, 23.
Semi-diurnal Arch, 67.
Continuance above the Horizon 9 Hours.

PROB. XLII. *The Latitude and the Day of the Month being given ; to find the Hour when any known Star will be on the Meridian, and alfo the Time of its Rifing and Setting.*

Having rectified the Globe for the Latitude and the Sun's Place; bring the given Star to the Meridian, and alfo to the Eaft or Weft Side of the Horizon, and the Index will fhew accordingly, when the Star *Culminates*, or the Time of its *Rifing* or *Setting*.

Thus at *London* on the 10th of *January*, *Syrius* will be upon the Meridian at a Quarter paft Ten in the Evening; rifes at 5¾ Hours, and fets at three Quarters paft two in the Morning.

By

By the Converfe of this Problem, knowing the Time when any Star is upon the Meridian, you may eafily find the *Sun's Place*.
Thus, bring the given Star to the Meridian,
and fet the Index to the given Hour ; then
turn the Globe until the Index points to 12
at Noon, and the Meridian will cut the *Sun's
Place* in the Ecliptick. Thus when *Syrius*
comes to the Meridian at 10¼ Hours after
Noon, the Sun's Place will be ♒ 1¼ Degrees.

PROB. XLIII. *To find at what Time of the
Year a given Star will be upon the Meridian, at a given Hour of the Night.*

Bring the Star to the Meridian, and fet the
Index to 12 at Noon, then turn the Globe
till the Index points to the given Hour, and
the Meridian will cut the Ecliptick in the
Sun's Place ; whence the Day of the Month
may be eafily found in the *Kalendar* upon
the Horizon.

PROB. XLIV. *The Day of the Month, and
the Azimuth of any known Star being given; to find the Hour of the Night.*

Having rectified the Globe for the Latitude
and the Sun's Place; if the given Star be
due North or South, bring it to the Meridian, and the Index will fhew the Hour of the
Night. If the Star be in any other Direction,
fix

fix the Quadrant of Altitude in the Zenith, and set it to the Star's Azimuth in the Horizon; then turn the Globe about until the Quadrant cuts the Center of the Star, and the Index will shew the *Hour* of the Night.

The Bearing of any Point in the Heavens may be found by the following Methods.

Having a Meridian Line drawn in two Windows, that are opposite to one another, you may cross it at right Angles with another Line, representing the East and West; from the Point of Intersection describe a Circle, and divide each Quadrant into 90 Degrees; then get a smooth Board, of about 2 Feet long, and $\frac{3}{4}$ Foot broad, (more or less as you judge convenient) and on the back part of it fix another small Board crosswise, so that it may serve as a Foot to support the biggest Board upright, when it is set upon a Level, or an Horizontal Plain. The Board being thus prepared, set the lower Edge of the smooth, or fore Side of it, close to the Center of the Circle, then turn it about to the Meridian, or to any Azimuth Point required, (keeping the Edge of it always close to the Center) and casting your Eye along the flat Side of it, you'll easily perceive what Stars are upon the Meridian, or any other Bearing that the Board is set to.

P R O B.

PROB. XLV. *Two known Stars having the same Azimuth, or the same Height, being given; to find the Hour of the Night.*

Rectify the Globe for the Latitude, the Zenith, and the Sun's Place.

1. When the two Stars are in the same Azimuth. Turn the Globe, and also the Quadrant about, until both the Stars coincide with the Edge thereof; then will the Index shew the Hour of the Night: and where the Quadrant cuts the Horizon, is the common Azimuth of both Stars.

2. If the two Stars are of the same Altitude, move the Globe so, that the same Degree on the Quadrant will cut both Stars; then the Index will shew the Hour.

This Problem is useful when the Quantity of the Azimuth of the two Stars, in the first Case, or of their Altitude in the latter Case, is not known.

If two Stars were given, one on the Meridian, and the other in the East or West Part of the Horizon; to find the Latitude.

Bring that Star, which was observed on the Meridian, to the Meridian of the Globe, and keep the Globe from turning round its Axis; then slide the Meridian up or down

in

in the Notches, till the other Star is brought
to the Eaſt or Weſt Part of the Horizon,
and that Elevation of the Pole will be the
Latitude ſought.

PROB. XLVI. *The Latitude, Day of*
the Month, and the Altitude of any
known Star being given; to find the
Hour of the Night.

Rectify the Globe for the Latitude, Zenith,
and Sun's Place: Turn the Globe, and the
Quadrant of Altitude, backward or forward,
till the Center of that Star meets the Quadrant in the Degree of Altitude given;
then the Index will point the true Hour of
the Night; and alſo where the Quadrant
cuts the Horizon, will be the Azimuth of
the Star at that Time.

If the Latitude, Sun's Altitude, and his
Declination (inſtead of his Place in the
Ecliptick) are given; to find the Hour
of the Day, and Azimuth.

Rectify the Globe for the Latitude and
Zenith; and having brought the *Equinoctial*
Colure to the Meridian, ſet the Index to 12
at Noon: which being done, turn the Globe
and the Quadrant, until the given Declination in the Equinoctial Colure cuts the Altitude on the Quadrant; then the Index

L   will

will fhew the *Hour* of the Day, and the Quadrant cut the *Azimuth* in the Horizon.

If the Altitude of two Stars on the fame Azimuth, were given; to find the La-titude of the Place.

Set the Quadrant over both Stars at the obferved Degree of Altitude, and keep it faft upon the Globe with your Fingers; then flide the Meridian up or down in the Notches, till the Quadrant cuts the given Azimuth in the Horizon; that Ele-vation of the Pole will be the Latitude re-quired.

PROB. XLVII. *Having the Latitude of the Place, to find the Degree of the Ecliptick, which rifes or fets with a given Star; and from thence to deter-mine the Time of its* Cofmical *and* Achro-nical *Rifing and Setting.*

Having rectified the Globe for the Lati-tude, bring the given Star to the Eaftern Side of the Horizon, and mark what De-gree of the Ecliptick rifes with it: Look for that Degree in the Wooden Horizon, and right againft it, in the Kalendar, you'll find the Month and Day when the Star *rifes Cofmically.* If you bring the Star to the Weftern Side of the Horizon, that Degree

of

of the Ecliptick which rifes at that Time, will give the Day of the Month, when the faid Star *fets Cofmically*. So likewife againft the Degree which fets with the Star, you'll find the Day of the Month of the *Achronical Setting* ; and if you bring it to the Eaftern Part of the Horizon, that Degree which fets at that Time will be the Sun's Place when the Star *rifes Achronically.*

Thus, in the Latitude of *London, Syrius,* or the *Dog-Star,* rifes *Cofmically* the 30th of *July* ; and fets *Cofmically* the 30th of *Octuber. Aldebaran,* or the *Bull's-Eye,* rifes *Achronically* on the 11th of *May,* and fets *Achronically* on the 8th of *December.*

PROB. XLVIII. *Having the Latitude of the Place, to find the Time when a Star rifes and fets* Heliacally.

Having rectified the Globe for the Lati tude, bring the Star to the Eaftern Side of the Horizon, and turn the Quadrant round to the Weftern Side, till it cuts the Ecliptick in twelve Degrees of Altitude above the Horizon, if the Star be of the firft Magni- tude ; then that Point of the Ecliptick which is cut by the Quadrant, is 12 Degrees high, above the Weftern Part of the Horizon, when the Star rifes ; but at the fame time the oppoffte Point in the Ecliptick is 12

L 2   De-

Degrees below the Eaſtern Part of the Horizon, which is the Depreſſion of a Star of
the *firſt Magnitude*, when ſhe *riſes Heliacally* ; or has got ſo far from the Sun's Beams,
that ſhe may be ſeen in the Morning before
Sun riſing. Wherefore look for the ſaid
Point of the Ecliptick on the Horizon, and
right againſt it will be the Day of the Month
when the Star *riſes Heliacally*. To find the
Heliacal Setting : Bring the Star to the
Weſt Side of the Horizon, and turn the
Quadrant about to the Eaſtern Side, till the
12*th* Degree of it above the Horizon, cuts
the Ecliptick ; then that Degree of the E-
cliptick, which is oppoſite to this Point, is
the Sun's Place when the Star *ſets Heliacally*.

Thus, you'll find that *Arcturus* riſes Heliacally the 17*th* of *September* ; and ſets
Heliacally, *November* the 21*ſt*.

PROB. XLIX. *To find the Place of any
Planet upon the Globe ; and ſo by that
means, to find its Place in the Heavens :
Alſo to find at what Hour any Planet
will riſe or ſet, or be on the Meridian,
at any Day in the Year.*

You muſt firſt ſeek in an Ephemerides,
(*Parker*'s Ephemeris will do well enough)
for the Place of the Planet propoſed on that
Day ; then mark that Point of the Ecliptick,
either

either with Chalk, or by flicking on a little black Patch ; and then for that Night you may perform any Problem, as before, by a Fixed Star.

Let it be required to find the Situation of *Jupiter* among the Fixed Stars in the Heavens, and alfo whenabouts it rifes and fets, and comes to the Meridian, on the 20*th* of *November*, 1730, at *London.*

Looking for the 20*th* of *November* 1730, in *Parker's Ephemeris,* I find that *Jupiter's* Place at that Time is in about 9 Degrees of ♍; Latitude 1 Degree North. Then looking for that Point upon the Celeftial Globe, I find that ♃ is then among the fmall Stars that lie under the Belly of the Conftellation *Leo.*

To find when he rifes and fets, and comes to the Meridian : Having put a little black Patch on the Place of *Jupiter,* elevate the Globe according to the Latitude; and having brought the Sun's Place to the Meridian, fet the Hour-Index to 12 at Noon : then turning the Mark which was made for *Ju- piter,* to the Eaftern Part of the Horizon, I find ♃ will rife at about a quarter paft 11 o'Clock at Night; and turning the Globe about, I find it comes to the Meridian a little after Six in the Morning ; and fets about 1 o'Clock in the Afternoon.

This Example being underftood, it will be eafy to find when either of the other two fuperior Planets, *viz. Mars* and *Sa-turn*, Rife, Set, and come to the Meridian.

I fhall conclude this Subject about the Globes with the following Problem.

PROB. L. *To find all that Space upon the Earth, where an Eclipfe of one of the Satellits of* Jupiter *will be vifible.*

Having found that Place upon the Earth, in which the Sun is Vertical, at the Time of the Eclipfe, by *Prob.* 13. Elevate the Globe according to the Latitude of the faid Place; then bring the Place to the Meridian, and fet the Hour-Index to 12 at Noon. If *Jupiter* be in Confequence of the Sun, draw a Line with Black-Lead, or the like, along the Eaftern Side of the Horizon, which Line will pafs over all thofe Places where the Sun is fetting at that Time: then count the Difference betwixt the Right Afcenfion of the Sun, and that of *Jupiter*; and turn the Globe Weftward, until the Hour-Index points to this Difference; then keep the Globe from turning round its Axis, and elevate the Meridian, according to the Declination of *Ju-piter*. The Globe being in this Pofition, draw a Line along the Eaftern Side of the Horizon, then the Space between this Line, and the Line befure drawn, will compre-hend all thofe Places of the Earth where *Jupiter* will be vifible, from the fetting of the Sun, to the fetting of *Jupiter*.

But

But if *Jupiter* be in Antecedence of the Sun, (*i.e.* rifes before him) having brought the Place where the Sun is Vertical, to the Zenith, and put the Hour-Index to 12 at Noon, draw a Line on the Weftern Side of the Horizon; then elevate the Globe according to the Declination of *Jupiter*, and turn it about Eaftwards, until the Index points to fo many Hours diftant from Noon, as is the Difference of Right Afcenfion of the Sun and *Jupiter*. The Globe being in this Pofition, draw a Line along the Weftern Side of the Horizon; then the Space contained between this Line, and the other laft drawn, will comprehend all thofe Places upon the Earth, where the Eclipfe is vifible, between the rifing of the Sun and *Jupiter*.

L 4 *The*

The DESCRIPTION *of the* Great ORRERY, *lately made by* Mr. THO. WRIGHT, Mathematical Inftrument-Maker *to His* MAJESTY.

Orrery. HE ORRERY is an Aftronomical Machine, made to reprefent the Motions of the Planets, as they really are in Nature: Thefe Machines are made of various Sizes, fome having more Planets than others; but I fhall here confine my felf to the Defcription of that above mentioned.

In the Introduction we gave a fhort Account of the *Order, Periods, Diftances,* and *Magnitudes* of the *Primary Planets;* and of the *Diftances* and *periodical Revolutions* of the *Secondary Planets* round their refpective Primaries. We fhall here explain their *Stations, Retrogradations, Eclipfes,* and *Phafes, &c.* but firft let us take a general view of the *Orrery.*

The

The Frame which contains the Wheel- *The De-scription* Work, *&c.* that regulates the whole *Ma-* *of the Or-chine,* is made of fine Ebony, and is near *rery.* four Feet in Diameter ; the outfide thereof is adorned with twelve Pilafters, curioufly wrought and gilt : Between thefe Pilafters, the twelve Signs of the *Zodiack* are neatly painted, within gilded Frames.   Above the Frame is a broad Ring, fupported with twelve *vide Fron-* Pillars : This Ring reprefents the *Plane* of *tifpiece.* the *Ecliptick,* upon which there is two Scales of Degrees, and between thofe the Names and Characters of the twelve Signs. Near the Outfide is a Scale of Months and Days, exactly correfponding to the Sun's Place at Noon, each Day throughout the Year.

Above the Ecliptick ftand fome of the principal Circles of the Sphere, according to their refpective Situation in the Heavens, *viz.* N° 10, are the two *Colures,* divided into Degrees, and Half Degrees : N° 11, is one Half of the Equinoctial Circle, making an Angle with the Ecliptick of $23\frac{1}{2}$ Degrees : The *Tropick* of *Cancer,* and the *Arctick Circle,* are each fixed parallel, and at their proper Diftance from the Equinoctial.   On the Northern Half of the Ecliptick, is a brafs Semicircle, moveable upon two Points fixed in ♈ and ♎ : This Semicircle ferves as a moveable Horizon, to be put to any De-gree of Latitude upon the North Part of
the

the Meridian. The whole Machine is also so contrived, as to be set to any Latitude, without in the least affecting any of the inside Motions : For this purpose there are two strong Hinges (N° 13.) fixed to the bottom Frame, upon which the Instrument moves, and a strong Brass Arch, having Holes at every Degree, thro' which a strong Pin is to be put, according to the Elevation. This Arch, and the two Hinges, support the whole Machine, when it is lifted up according to any Latitude ; and the Arch at other times lies conveniently under the bottom Frame.

When the Machine is set to any Latitude (which is easily done, by two Men, each taking hold of two Handles, conveniently fixed for that purpose) set the moveable Horizon to the same Degree upon the Meridian, and you may form an Idea of the respective Altitude, or Depressions of the Planets, above or below the Horizon, according to their respective Positions, with regard to the Meridian.

Within the Ecliptick, and nearly in the same Place thereof, stand the Sun, and all the Planets both Primary and Secondary. The Sun (N° 1.) stands in the middle of the whole System, upon a Wire, making an Angle with the Plane of the Ecliptick, of about 82 Degrees, which is the Inclination of the Sun's Axis, to the Axis of the Ecliptick. Next the Sun is a small Ball, (N° 2.) repre-

representing *Mercury* : Next to *Mercury*
is *Venus* (N° 3.) reprefented by a larger
Ball, (and both thefe ftand upon Wires, fo
that the Balls themfelves may be more vifi-
bly perceived by the Eye.)   The Earth is reprefented (N° 4.) by an Ivory Ball, having
fome of the principal Meridians and Parallels, and a little Sketch of a Map defcribed
upon it.   The Wire which fupports the
Earth, makes an Angle with the Plane of the
Ecliptick of 66½ Degrees, which is the Inclination of the Earth's Axis to that of the
Ecliptick.   Near the Bottom of the Earth's
Axis is a Dial Plate, (N° 9.) having an Index
pointing to the Hours of the Day, as the
Earth turns round its Axis.

Round the Earth is a Ring fupported by
two fmall Pillars; which Ring reprefents
the Orbit of the Moon, and the Divifions
upon it anfwer to the Moon's Latitude ; the
Motion of this Ring reprefents the Motion
of the Moon's Orbit according to that of
the Nodes.   Within this Ring is the Moon
(N° 5.) having a black Cap or Cafe, which
by its Motion reprefents the *Phafes* of the
Moon according to her Age.   Without the
Orbits of the Earth and Moon is *Mars*
(N° 6.) The next in Order to *Mars*, is *Ju-*
piter and his four Moons, (N° 7.); each of
thefe Moons is fupported by a crooked
Wire, fixed in a Socket which turns about
the Pillar that fupports *Jupiter :*   thefe Satellits

tellits may be turned by the Hand to any Position; and yet when the Machine is put in motion, they'll all move in their proper Times. The outermost of all is *Saturn* and his five Moons, (N° 8.) these Moons are supported and contrived after the same manner with those of *Jupiter*. The whole Machine is put into Motion by turning a small Winch, (like the Key of a Clock, N° 14); and all the inside Work is so truly wrought, that notwithstanding the vast Number of Wheels that are to be turned, it does not require above the Strength of a fine Hair to put the whole in motion.

Above the Handle, there is a Cylindrical Pin, which may be drawn a little out, or pushed in at pleasure : When it is pushed in, all the Planets both Primary and Secondary will move according to their respective Periods, by turning the Handle : when it is drawn out, the Motions of the Satellits of *Jupiter* and *Saturn* will be stopped, while all the rest move without interruption. This is a very good Contrivance to preserve the Instrument from being clogged by the swift Motions of the Wheels belonging to the Satellits of *Jupiter* and *Saturn*, when the Motions of the rest of the Planets are only considered.

There is also a brass Lamp, having two convex Glasses, to be put in the room of the Sun; and also a smaller Earth and Moon,

2   made

made fomewhat in Proportion to their Diftance from each other, which may be put
on at pleafure.

The Lamp turns round in the fame time
with the Earth, and by means of the Glaffes
cafts a ftrong Light upon her: and when the
fmaller Earth and Moon are placed on, it
will be eafy to fhew when either of them
may be eclipfed.

Having thus given a brief Defcription of
the outward Parts of this Machine, I fhall
next give an Account of the Phenomena
explained by it when it is put into Motion.

I. *Of the Motions of the Planets in general.*

Having put on the Handle, pufh in the
Pin which is juft above it, and place a fmall
black Patch (or bit of Wafer) upon the middle of the Sun (for Inftance) right againft
the firft Degree of ♈; you may alfo place
Patches upon *Venus, Mars* and *Jupiter,*
right againft fome noted Point in the Ecliptick. If you lay a Thread from the Sun to
the firft Degree of ♈, you may fet a Mark
where it interfects the Orbit of each Planet ;
and that will be a help to note the Time of
their Revolutions.

One entire Turn of the Handle anfwers
to the Diurnal Motion of the Earth round
her Axis; as may be feen by the Motion of
the Hour Index which is placed at the Foot of
the

the Wire on which the Terella is fixed.
When the Index has moved the fpace of ten
Hours, you may obferve that *Jupiter* has
made one Revolution compleat round its
Axis; the Handle being turned until the
Hour-Index has paffed over 23 Hours, will
bring the Patch upon *Venus* to its former
Situation with refpect to the Ecliptick,
which fhews that ♀ has made one entire Revolution round her Axis. *Mars* makes one
compleat Revolution round his Axis in 24
Hours and about 40 Minutes. When the
Handle is turned $25\frac{1}{2}$ times round, the Spot
upon the Sun will point to the fame Degree
of the Ecliptick, as it did when the Inftrument was nrft put into motion. By obferving
the Motions of the Spots upon the Surface
of the Sun, and of the Planets in the
Heavens, their Diurnal Motion was difcovered; after the fame manner as we do
here obferve the Motions of their Repre-
fentatives, by that of the Marks placed upon them.

If while you turn the Handle you obferve
the Planets, you will fee them perform their
Motions in the fame relative Times as they
really do in the Heavens; each making its
Period in the Times mentioned in the Ta-
bles, Page 24. $27\frac{1}{4}$ Turns of the Handle will
bring the Moon round the Earth, which is
called a *Periodic* Month, and all the while
fhe keeps the fame Face towards the Earth:
for

for the Moon's Annual and Diurnal Motion
are performed both in the fame Time nearly, fo that we always fee the fame Face or
Side of the Moon.

If before the Inftrument is put into Motion, the Satellits of *Jupiter* and *Saturn*
be brought into the fame right Line from
their refpective Primaries, you'll fee them
as you turn the Handle immediately difpers'd
from one another, according to their different Celerities. Thus, one Turn of the
Handle will bring the firft of *Jupiter*'s
Moon's about $\frac{4}{7}$ part round *Jupiter*; while
the fecond has defcribed but $\frac{2}{7}$ part; the
third but about $\frac{1}{7}$; and the fourth not quite
$\frac{1}{18}$ part; each of its refpective Orbit. If
you turn the Handle until the Hour-Index
has moved $18\frac{1}{4}$ Hours more; the firft Satel-
lit will be then brought into its former Pofition, and fo has made one entire Revolution; the fecond at the fame time will be
almoft diametrically oppofite to the firft;
and fo has made a little more than half of
one Revolution: the others will be in different Afpects, according to the Length of
their Periods, as will be plainly exhibited
by the Inftrument. The fame Obfervations
may be made with refpect to the Satellits
of *Saturn*.

The Machine is fo contrived, that the
Handle may be turned either way; and if
before you put it into motion, you obferve

the

the Aspect (or Situation with respect to each
other) of the Planets, and then turn the
Handle round any Number of Times; the
same number of Revolutions being made
backwards, will bring all the Planets to their
former Situations. I shall next proceed 'to
Particulars.

Of the Stations and Retrogradations of the Planets.

The Primary Planets, as they all turn
round the Sun, at different Distances and
in different Times, appear to us from the
Earth to have different Motions; as sometimes they appear to move from West to
East according to the Order of the Signs,
which is called their *direct* Motion; then by
degrees they slacken their pace, until at last
Stationary. they lose all their Motion and become *Sta-*
tionary, or not to move at all; that is, they
appear in the same place with respect to the
fixed Stars for some time together: After
which they again begin to move, but with a
contrary Direction, as from East to West;
Retrograde which is called their *retrograde* Motion:
Motion of then again they become Stationary, and af-
the Planets. terwards reassume their direct Motion. The
Reason of all these Appearances is very evi-
dently shewn by the *Orrery*.

I. *Of*

Of the Stations, &c. *of the inferior Planets.*

We fhall inftance in the Planet *Mercury*, becaufe his Motion round the Sun differs more from the Earth's than that of *Venus* does.

When *Mercury* is in his fuperior Conjunction (or when he is in a direct Line from the Earth beyond the Sun) faften a String about the Axis of the Earth, and extend it over *Mercury* to the Ecliptick; then turning the Handle, keep the Thread all the while extended over ☿, and you'll find it move with a direct Motion in the Ecliptick, but continually flower, until *Mercury* has the greateft Elongation from the Earth. Near this Pofition, the Thread for fome time will lay over *Mercury* without being moved in the Ecliptick, tho' the Earth and *Mercury* both continue their progreffive Motion in their refpective Orbits. When *Mercury* has got a little paft this Place, you'll find the Thread muft be moved backward in the Ecliptick, beginning firft with a flow Motion, and then fafter by degrees, until *Mercury* is in his inferior Conjunction, or directly betwixt the Earth and the Sun. Near this Pofition of ☿, his retrograde Motion will be the fwifteft; but he ftill moves the fame way, tho' continually flower, till he has again

M   come

come to his greatest Elongation, where he
will appear the second time to be Stationary;
after which, he begins to move forward,
and that faster by degrees, until he is come
to the same Position, with respect to the
Earth, that he was in at first. The same
Observations may be made relating to the
Motions of *Venus*. In like manner the
different Motions observed in the superior
Planets, may be also explained by the *Orrery*.
If you extend the Thread over *Jupiter*, and
proceed after the same manner as before we
did in regard to *Mercury*; you'll find that
from the time *Jupiter* is in Conjunction
with the Sun, his Motion is direct, but continually slower until the Earth is nearly in a
Quadrate Aspect with *Jupiter*; near which Position *Jupiter* seems to be Stationary: after
which he begins to move, and so continually
mends his Pace until he comes in opposition to the Sun, at which time his retrograde Motion is swiftest. He still seems to go
backward, but with a slower Pace, till the
Earth and he are again in a Quadrate Aspect,
where *Jupiter* seems to have lost all his
motion; after which he again resumes his direct Motion, and so proceeds faster by degrees, till the Earth and he are again in opposition to each other.

   These different Motions observed in the
Planets, are easily illustrated as followeth:
Plate 3. The lesser Circle round the Sun is the Orbit
Fig. 1.   3   of

of *Mercury*, in which he performs his Revolution round the Sun in about three
Months, or while the Earth is going through
¼ part of her Orbit, or from A to N. The
Numbers 1, 2, 3, &c. in the Orbit of *Mercury*, shew the Spaces he describes in a Week
nearly ; and the Distances AB, BC, CD, &c.
in the Earth's Orbit, do likewise shew her
Motion in the same Time. The Letters
A, B, C, &c. in the great Orb, are the
Motions of *Mercury* in the Heavens, as they
appear from the Earth. Now if the Earth
be supposed in A, and *Mercury* in 12, near
his superior Conjunction with the Sun ; a
Spectator on the Earth will see ☿, as if
he were in the Point of the Heavens A ; and
while ☿ is moving from 12 to 1, and from
1 to 2, &c. the Earth in the same time also
moves from A to B, and from B to C, &c.
All which time ☿ appears in the Heavens to
move in a direct Motion from A to B, and
from B to C, &c. but gradually slower until he arrives near the Point G, near which
place he appears Stationary, or to stand still :
and afterwards (tho' he still continues to
move uniformly in his own Orbit with a
progressive Motion) yet in the Sphere of the
fixed Stars he'll appear to be retrograde, or
to go backwards, as from G to H, from H to
I, &c. until he has arrived near the Point L,
where again he'll appear to be Stationary, and

M 2   af-

afterwards to move in a direct Motion from
L to M, and from M to N, &c.

What has been here fhewed concerning
the Motions of *Mercury*, is alfo to be underftood of the Motions of *Venus* ; but the
Conjunctions of *Venus* with the Sun, do not
happen fo often as in *Mercury* : for *Venus*
moving in a larger Orbit, and much flower
than *Mercury*, does not fo often overtake
the Earth. But the Retrogradations are
much greater in *Venus* than they are in
Mercury, for the fame Reafons.

Fig. 2. The innermoft Circle reprefents the
Earth's Orbit, divided into 12 Parts, anfwer-
ing to her Monthly Motion : the greateft
Circle is the Orbit of *Jupiter*, which he
defcribes in about 12 Years ; and therefore
the $\frac{1}{12}$ thereof, from A to N, defines his Motion in one of our Years nearly ; and the
intermediate Divifions, A, B, C, &c. his
Monthly Motion. Let us fuppofe the Earth
to be in the Point of her Orbit 12, and
Jupiter in A, in his Conjunction with the
Sun : It is evident that from the Earth, *Jupiter* will be feen in the Great Orb, or in
the Point of the Heavens A ; and while the
Earth is moving from 12 to 1, 2, &c. ♃ alfo moves from A to B, C, &c. all which
Time he appears in the Heavens to move with
a direct Motion from A to B, C, &c. until
he comes in Oppofition to the Earth near the
Point of the Heavens E, where he appears

to

to be Stationary : After which, ♃ again begins to move (tho' at firſt with a ſlow pace)
from E thro' F, H, I, to K ; where again he appears to ſtand ſtill; but afterwards he reaſſumes
his direct motion from I thro' K, to M, &c.

From the Conſtruction of the preceding
Figure, it appears that when the Superior
Planets are in Conjunction with the Sun,
their direct Motion is much quicker than at
other times ; and that becauſe they really
move from Weſt to Eaſt, while the Earth
in the oppoſite Part of the Heavens is carried the ſame Way, and round the ſame Center. This Motion afterwards continually
ſlackens until the Planet comes almoſt in
oppoſition to the Sun, when the Line joining the Earth and Planet will continue for
ſome time nearly parallel to it ſelf, and ſo
the Planet ſeems from the Earth to ſtand ſtill ;
after which, it begins to move with a
ſlow Motion backward, until it comes
into a Quartile Aſpect with the Sun, when
again it will appear to be ſtationary for the
above Reaſons. After that, it will reſume
its direct Motion until it comes into a Conjunction with the Sun, then it will proceed
as above explained. Hence, it alſo appears
that the Retrogradations of the ſuperior
Planets, are much ſlower than their direct
Motions, and their continuance much ſhort-
er; for the Planet, from its laſt Quarter,
until it comes in oppoſition to the Sun, ap-

pears

pears to move the fame way with the Earth, by whom it is then overtaken: After which it begins to go backwards, but with a flow Motion, becaufe the Earth being in the fame part of the Heavens, and moving the fame way, that the Planet really does, the apparent Motion of the Planet backwards muft thereby be leffened.

What has been here faid concerning the Motions of *Jupiter*, is alfo to be underftood of *Mars* and *Saturn*. But the Retrogradations of *Saturn* do oftner happen than thofe of *Jupiter*, becaufe the Earth ofner overtakes *Saturn*; and for the fame Reafon, the Regreffions of *Jupiter* do oftner happen than thofe of *Mars*. But the Retrogradations of *Mars* are much greater than thofe of *Jupiter*; whofe are alfo much greater than thofe of *Saturn*.

In either of the Satellits of *Jupiter* or *Saturn*, thefe different Appearances in the neighbouring Worlds, are much oftner feen than we do in the Primary Planets.

We never obferve thefe different Motions in the Moon, becaufe fhe turns round the Earth as her Center; neither do we obferve them in the Sun, becaufe he is the Center of the Earth's Motion, whence the apparent Motion of the Sun always appears the fame way round the Earth,

Of

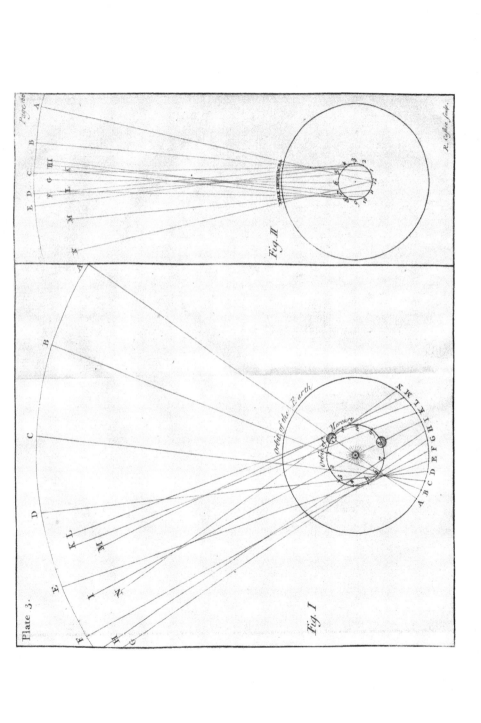

Plate 3.

Fig. II

Fig. I

orbit of the Earth

Mercury

R. Caslon Sculp.

Page 166

A B C D E F G H I K L M N

Of the *Annual* and *Diurnal Motion* of the *Earth,* and of the *Increase* and *Decrease* of *Days and Nights.*

The Earth in her Annual Motion round the Sun, has her Axis always in the same Direction, or parallel to it self: that is, if a Line be drawn parallel to the Axis while the Earth is in any point of her Orbit, the Axis in all other Positions of the Earth, will be parallel to the said Line This Parallelism of the Axis, and the simple Motion of the Earth in the Ecliptick, solves all the Phenomena of different Seasons. These things are very well illustrated by the *Orrery.*

If you put on the Lamp in the place of the Sun, you will see how one half of our Globe is always illuminated by the Sun, while the other Hemisphere remains in darkness ; how Day and Night are formed by the Revolution of the Earth round her Axis : for as she turns from West to East, the Sun appears to move from East to West. And while the Earth turns in her Orbit, you may observe that her Axis always points the same way; and the several Seasons of the Year continually change.

To make these things plainer, we will take a View of the Earth in different parts of her Orbit.

When

When the Earth is in the firft Point of *Libra* (which is found by extending a Thread from the Sun, and over the Earth to the Ecliptick) we have the *Vernal Equinox*, and the Sun at that Time appears in the firft Point of ♈. In this Pofition of the Earth, the two Poles of the World are in the Line feparating Light and Darkness ; and as the Earth turns round her Axis, juft one Half of the Equator, and all its Parallels, will be in the Light, and the other Half in the Dark; and therefore the Days and Nights muft be every where equal.

As the Earth moves along in her Orbit, you'll perceive the North Pole advances by degrees into the illuminated Hemifphere, and at the fame Time, the South Pole recedes into Darkness ; and in all Places to the Northward of the Equator, the Days continually lengthen, while the contrary happens in the Southern Parts; until at length the Earth is arrived in *Capricorn*. In this Pofition of the Earth, all the Space included within the Arctick Circle falls wholly within the Light ; and all the oppofite Part lying within the Antarctick Circle, is quite involved in Darkness. In all Places between the Equator and the Arctick Circle, the Days are now at the longeft, and are gradually longer as the Places are more remote from the Equator : In the Southern Hemifphere there is the contrary

trary

trary Effect. All the while the Earth is travelling from *Capricorn* towards *Aries*, the
North Pole gradually recedes from the Light,
and the South Pole approaches nearer to it:
the Days in the Northern Hemisphere gradually decrease, and in the Southern Hemisphere they increase in the same proportion,
until the Earth being arrived in ♈, then the
two Poles of the World lie exactly in the
Line separating Light and Darkness, and
the Days are equal to the Nights in all
Places of the World. As the Earth advances
towards *Cancer*, the North Pole gradually
recedes from the Light, while the Southern
one advances into it, at the same rate; in
the Northern Hemisphere the Days decrease,
and in the Southern ones they gradually
lengthen, until the Earth being arrived in
Cancer, the North Frigid Zone is all involved in Darkness, and the South Frigid
Zone falls entirely within the Light; the
Days every where in the Northern Hemisphere are now at the shortest, and to the
Southward they are at the longest. As the
Earth moves from hence towards *Libra*, the
North Pole gradually approaches the Light,
and the other recedes from it; and in all
Places to the Northward of the Equator,
the Days now lengthen, while in the op-
posite Hemisphere they gradually shorten,
until the Earth has got into ♎; in which
Position, the Days and Nights will again
be

be of equal Length in all Parts of the World.

You might have obſerved, that in all Poſitions of the Earth, one half of the Equator was in the Light, and the other half in Darkneſs; whence under theEquator,theDays and Nights are always of the ſame Length: And all the while the Earth was going from ♎ towards ♈, the North Pole was conſtantly illuminated, and the South Pole all the while in Darkneſs; and for the other Half Year the contrary. Sometimes there is a Semicircle exactly facing the Sun, fixed over the Middle of the Earth, which may be called the Horizon of the Disk : This will do inſtead of the Lamp, if that Half of the Earth which is next the Sun be conſider'd, as being the illuminated Hemiſphere, and the other Half, to be that which lies in Darkneſs.

Plate 4. The Great Circle ♈, ♉, ♊, &c. repreſents the Earth's annual Orbit; and the four leſſer Circles E S Q C, the Ecliptick upon the Surface of the Earth, coinciding with the great Ecliptick in the Heavens. Theſe four leſſer Figures repreſent the Earth in the four Cardinal Points of the Ecliptick; P being the North Pole of the Equator; and *p* the North Pole of the Ecliptick; S P C the Solſtitial Colure, which is always parallel to the great Solſtitial Colure ♋ ☉ ♑ in the Heavens; E P Q the Equinoctial Colure. The other
Circles

Circles paffing thro' P, are Meridians at two Hours diftance from one another; the Semi-circle E Æ Q is the Northern Half of the Equator; the parallel Circle touching the Ecliptick in S, is the Tropick of *Cancer*; the dotted Circle the Parallel of *London*; and the fmall Circle, touching the Pole of the Ecliptick, is the *Arctick Circle*. The fhaded Part, which is always oppofite to the Sun, is the obfcure Hemifphere, or that which lies in Darknefs; and that which is next the Sun, is the illuminated Hemifphere.

If we fuppofe the Earth in ♎, fhe'll then fee the Sun in ♈, (which makes our vernal Equinox) and in this Pofition, the Circle, bounding Light and Darknefs, which here is S C, paffes thro' the Poles of the World, and bifects all the Parallels of the Equator; and therefore the Diurnal and Nocturnal Arches, or the Lengths of the Days and Nights are equal in all Places of the World.

But while the Earth, in her annual Courfe, moves thro' ♏, ♐, to ♑; the Line S C, keep-ing ftill parallel to it felf, or to the Place where it was at firft; the Pole P, will by this Motion gradually advance into the illu-minated Hemifphere; and alfo the Diurnal Arches of the Parallels gradually increafe, and confequently the Nocturnal ones de-creafe in the fame proportion, until the Earth has arrived into ♑ : in which Pofition the Pole P, and all the Space within the

i   Arctick

Arctick Circle, fall wholly within the illuminated Hemisphere; and the Diurnal Arches of all the Parallels that are without this Circle, will exceed the Nocturnal Arches more or less, as the Places are nearer to, or farther off from it; until the Distance from the Pole is as far as the Equator, where both these Arches are always equal.

Again, while the Earth is moving from ♑, through ♒, ♓, to ♈; the Pole P begins to incline to the Line, distinguishing Light and Darkness, in the same proportion that before it receded from it; and consequently the Diurnal Arches gradually lessen, until the Earth has arrived into ♈; where the Pole P will again fall in the Horizon, and so cause the Days and Nights to be every where equal. But when the Earth has passed ♈, while she is going thro' ♉ and ♊, &c. the Pole P, will begin to fall in the obscure Hemisphere, and so recede gradually from the Light, until the Earth is arrived in ♋; in which Position, not only the Pole, but all the Space within the Arctick Circle, are involved in Darkness; and the Diurnal Arches of all the Parallels, without the Arctick Circle, are equal to the Nocturnal Arches of the same Parallels, when the Earth was in the opposite Point ♑: and it is evident, that the Days are now at the shortest, and the Nights the longest. But when the Earth has past this Point, while she is going

A   through

Plate 4

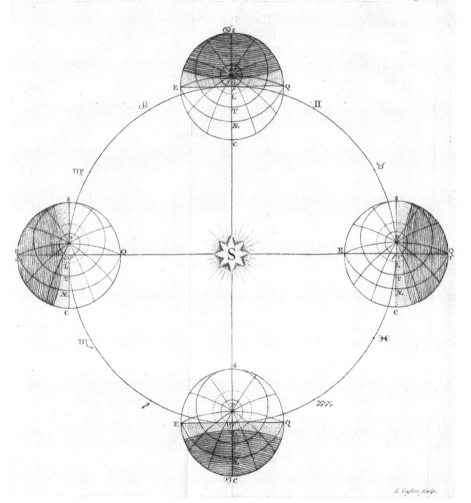

R. Cushee sculp.

through ♌, and ♍, the Pole P will again gradually approach the Light, and so the diurnal Arches of the Parallels gradually lengthen, until the Earth is arrived in ♎; at which Time the Days and Nights will again be equal in all Places of the World, and the Pole it self just see the Sun.

Here we only confidered the Phenomena belonging to the Northern Parallels; but if the Pole P be made the South Pole, then all the Parallels of Latitude will be Parallels of South Latitude; and the Days every where, in any Pofition of the Earth, will be equal to the Nights of thofe who lived in the oppofite Hemifphere, under the fame Parallels.

Of the Phafes of the Moon, and of her Motion in her Orbit.

The Orbit of the Moon makes an Angle with the Plane of the Ecliptick, of above $5\frac{1}{4}$ Degrees, and cuts it into two Points, diametrically oppofite, (after the fame manner as the Equator and the Ecliptick cut each other upon the Globe in ♈ and ♎) which Points are called the *Nodes* ; and a right Line *Nodes.* joining thefe Points, and paffing through the Center of the Earth, is called the *Line* Line of the of the Nodes. That Node where the Moon *Nodes.* begins to afcend Northward above the Plane of the Ecliptick, is called the *Afcending Node*, and the *Head of the Dragon*, and Dragon's- is thus commonly marked ☊. The other *Head.* Node, from whence the Moon defcends to

2   the

the Southward of the Ecliptick, is called the
Dragon's- *Descending Node,* and the *Dragon's-Tail,*
Tail.   and is marked ☷. The Line of the Nodes
continually shifts it self from East to West,
contrary to the Order of the Signs; and
Retrograde with this *Retrograde* Motion, makes one
Motion of Revolution round the Earth, in the Space
the Nodes. of about 19 Years.

The Moon describes its Orbit round the
Earth in the Space of 27 Days, and 7 Hours,
Periodical which Space of Time is called a *Periodical*
Month.   *Month;* yet from one Conjunction to the
next, the Moon spends 29 Days and a Half,
Synodical. which is called a *Synodical Month;* because
Month.   while the Moon in its proper *Orbit* finishes
its Course, the Earth advances near a whole
Sign in the Ecliptick, which Space the Moon
has still to describe before she will be seen
in Conjunction with the Sun.

When the Moon is in Conjunction with
the Sun, note its Place in the Ecliptick;
then turning the Handle, you'll find that
27 Days and 7 Hours will bring the Moon
to the same Place; and after you have made
$2\frac{1}{4}$ Revolutions more, the Moon will be
exactly betwixt the Sun and the Earth.

The Moon all the while keeps in her
Orbit, and so the Wire that supports her
continually rises or falls in a Socket, as she
changes her Latitude; the black Cap shifts
Phases of it self, and so shews the Phases of the Moon,
the Moon. according to her Age, or how much of her
en-

enlighten'd Part is feen from the Earth. In one Synodical Month, the Line of the Nodes moves about $1\frac{1}{2}$ Degree from Weft to Eaft, and fo makes one entire Revolution in 19 Years.

Let A B be an Arch of the Earth's Orbit, *Plate 5.* and when the Earth is in T, let the Moon *Fig. 1.* be in N, in Conjunction with the Sun in S; while the Moon is defcribing her Orbit N A F D, the Earth will defcribe the Arch of her Orbit T *t*; and when the Earth has got into the Point *t*, the Moon will be in the Point of her Orbit *n*, having made one compleat Revolution round the Earth. But the Moon, before fhe comes in Conjunction with the Sun, muft again defcribe the Arch *n o*; which Arch is fimilar to T *t*, becaufe the Lines F N, *f n*, are parallel; and becaufe, while the Moon defcribes the Arch *n o*, the Earth advances forward in the Ecliptick; the Arch defcribed by the Moon, after fhe has finifhed her periodical Month, before fhe makes a Synodical Month, muft be fomewhat greater than *n o*. To determine the mean Length of a Synodical Month; find the Diurnal Motion of the Moon, (or the Space fhe defcribes round the Earth in one Day) and likewife the Diurnal Motion of the Earth; then the Difference betwixt thefe two Motions, is the apparent Motion of the Moon round the Earth in oneDay: then it will be, As this differential Arch,is to a whole

Circle;

Circle ; so is one Day, to that Space of Time wherein the Moon appears to describe a compleat Circle round the Earth ; which is about 29¼ Days : But this is not always a true *Lunation* ; for the Motion of the Moon is sometimes faster and sometimes slower, according to the Position of the Earth in her Orbit.

In one Synodical Month, the Moon has all manner of Aspects with the Sun and Earth ; and because she is opaque, that Face of hers will only appear bright which is towards the Sun, while the opposite remains in Darkness. But the Inhabitants of the Earth can only see that Face of the Moon which is turned towards the Earth ; and therefore, according to the various Positions of the Moon, in respect of the Sun and Earth, we observe different Portions of her illuminated Face, and so a continual Change in her * *Phases*.

Let S be the Sun, R T V an Arch of the Earth's Orbit, T the Earth, and the Circle A B C D, *&c.* the Moon's Orbit, in which she turns round the Earth, in the Space of a Month ; and let A, B, C, *&c.* be the Centers of the Moon in different Parts of her Orbit.

* *Phases* of the Moon, are those different Appearances we observe in her, according to her Position in respect of the Sun and Earth.

Now

Now if with the Lines S A, S B, *&c.*
we join the Centers of the Sun and Moon,
and at right Angles to thefe, draw the Lines
H O; the faid Lines H O, will be the Circles
that feparate the illuminated Part of the
Moon, from the dark and obfcure : Again, if
we conceive another Line I L, to be drawn
at right Angles to the Lines T A, T B, *&c.*
pafling from the Center of the Earth to the
Moon, the faid Line I L will divide the
vifible Hemifphere of the Moon, or that
which is turned towards us, from the invifible, or that which is turned from us;
and this Circle may be called the *Circle* of
Vifion.

Now it is manifeft, that whenever the *Full Moon.*
Moon is in the Pofition A, or in that Point
of her Orbit which is oppofite to the Sun,
the Circle of Vifion, and the Circle bounding Light and Darknefs do coincide, and
all the illuminated Face of the Moon is
turned towards the Earth, and is vifible to
us; and in this Pofition the *Moon* is faid to
be *full.* But when the Moon arrives to B,
all her illuminated Face is then not towards
the Earth, there being a Part of it, H B I,
not to be feen by us; and then her vifible
Face is deficient from a Circle, and appears
of a gibbous Form, as in B. *Fig.* 3. Again,
when fhe arrives to C, the two forementioned
Circles cut each other at right Angles, and
then we obferve a *Half Moon,* as in C, *Half Moon.*
     N   *Fig.*

Fig. 3. And again, the illuminated Face of the Moon is more and more turned from the Earth, until she comes to the Point E, where the Circle of Vision, and that bounding Light and Darkness, do again coincide. Here the Moon disappears, the illuminated Part being wholly turned from the Earth; and she is now said to be in *Conjunction* with the *Sun*, because she is in the same Direction from the Earth, that the Sun is New Moon. in, which Position we call a *New Moon*. When the Moon has arrived to F, she again resumes a horned Figure; but her Horns (which before the Change were turned Westward) have now changed their Position, and look Eastward. When she has arrived to a Quadrate Aspect at G, she'll appear bissected, like a Half-Moon; afterwards she'll still grow bigger, until at last she comes to A, where again she'll appear in her full Splendor.

The same Appearances which we observe in the Moon, are likewise observed by the *Lunarians* in the Earth; our Earth being a Moon to them, as their Moon is to us; and we are observed by them, to be carried round in the same Space of Time, that they are really carried round the Earth. But the same Phases of the Earth and Moon happen when they are in a contrary Position; for when the Moon is in Conjunction to us, the Earth is then in Opposition to the Moon, and the *Lunarians* have then a full Earth,

as

as we in a fimilar Pofition have a full Moon. When the Moon comes in Oppofition to the Sun, the Earth feen from the Moon, will appear in Conjunction with her, and in that Pofition the Earth will difappear; afterwards fhe'll affume a horned Figure, and fo fhew the fame Phafes to the Inhabitants of the Moon, as fhe does to us.

Of the Eclipfes of the Sun and Moon.

An *Eclipfe* is that Deprivation of Light *Eclipfe* in a Planet, when another is interpofed betwixt it and the Sun. Thus, an Eclipfe of the Sun is made by the Interpofition of the Moon at her Conjunction; and an Eclipfe of the Moon is occafioned by the Shadow of the Earth falling upon the Moon, when fhe is in Oppofition to the Sun.

Let S be the Sun, T the Earth, and *Fig 4* A B C its Shadow; now if the Moon, when fhe is in Oppofition to the Sun, fhould come into the conical Space A B C, fhe'll then be deprived of the Solar Light, and *Lunar* fo undergo an Eclipfe. *Eclipfe.*

In the fame manner when the Shadow of the Moon falls upon the Earth (which can never happen but when the Moon is in Conjunction with the Sun) that part upon which the Shadow falls, will be involved in Darknefs, and the Sun eclipfed. But becaufe the *Solar* Moon is much lefs than the Earth, the Sha- *Eclipfe.*

dow

Fig. 5.

dow of the ☽ cannot cover the whole Earth, but only a part of it. Let S be the Sun, T the Earth, A B C the Moon's Orbit, and L the Moon in Conjunction with the Sun : Here the Shadow of the Moon falls only upon the part DE of the Earth's Surface, and there only the Sun is entirely hid ; but there are other parts E F, D G, on each side of the Shadow, where the Inhabitants are deprived of part of the Solar Rays, and that more or less according to their Distance from the Shadow. Those who live at H and I, will see half of the Sun eclipsed : but in the Spaces F M. G N, all the Sun's Body will be visible without any Eclipse. From the preceding Figure, it appears, that an Eclipse of the Sun does not reach a great way upon the Superficies of the Earth : but the whole Body of the Moon may sometimes be involved in the Earth's Shadow.

Although the Moon seen from the Earth, and the Earth seen from the Moon, are each alternately once a Month in Conjunction with the Sun; yet by Reason of the Inclination of the Moon's Orbit to the Ecliptick, the Sun is not eclipsed every New Moon, nor the Moon at every Full. Let T be the Earth, D T E an Arch of the Ecliptick, A L B F the Moon's Orbit, having the Earth T in its Center ; and let A G B C be another Circle coinciding with the Ecliptick, and

Fig. 6.

A,

A, B, the Nodes, or the two Points where the Moon's Orbit and the Ecliptick cut each other; A the afcending Node, and B the defcending Node. The Angle G A L equal to G B L is the Inclination of the Moon's Orbit to the Ecliptick, being about $5\frac{1}{4}$ Degrees. Now a Spectator from the Earth at T, will obferve the Sun to move in the Circle A G B C, and the Moon in her Orbit A L B F; whence it is evident, that the Sun and Moon can never be feen in a direct Line from the Center of the Earth, but when the Moon is in one of the Nodes A or B; and then only will the Sun appear centrally eclipfed. But if the Conjunction of the Moon happens when fhe is any where within the Diftance A *c* of the Nodes either North or South, the Sun will be then eclipfed, more or lefs according to the Diftance from the Node A, or B. If the Conjunction happens when the Moon is in *b*, the Sun will be then one half eclipfed; and if it happens when fhe is in *c*, the Moon's Limb will juft touch the Sun's Disk, without hiding any part of it.

The Shadow of the Earth at the Place where the Moon's Orbit interfects it, is three times as large as the Moon's Diameter, as in *Fig.* 4. and therefore it often happens that Eclipfes of the Moon are Total, when they are not Central: and for the fame Reafon

N 3 the

the Moon may fometimes be totally eclipfed for three Hours together; whereas Total Eclipfes of the Sun can fcarcely ever exceed four Minutes.

The Eclipfes of the Sun and Moon are very well explained by the *Orrery* : thus, Having put the Lamp in the Place of the Sun, and the little Earth and the little Moon in their proper Places, inftead of the larger ones; let the Room wherein the Inftrument ftands be darkened; then turning the Handle about, you'll fee when the Conjunction of the Moon happens. When fhe is in or near one of the Nodes, her Shadow will fall upon the Earth, and fo deprive that part upon which it falls of the Light of the Sun : If the Conjunction happens when the Moon is not near one of the Nodes, the Light of the Lamp will fall upon the Earth, either above or below the Moon, according to her Latitude at that Time. In like manner, when the full Moon happens near one of the Nodes, the Shadow of the Earth will fall upon the Moon; and if the Moon's Latitude be but fmall, her whole Face will be involved in Darknefs. At other times, when the full Moon happens when fhe is not near one of her Nodes, the Shadow of the Earth will pafs either above or below the Moon, and fo by that means the Moon will efcape being eclipfed.

Of

Of the Eclipfes of the Satellits *of* Jupiter.

The apparent Diameters of the Inferior
Planets are fo fmall, that when they pafs
betwixt us and the Sun, they only appear
like fmall Spots upon the Sun's Surface,
without depriving us of any fenfible quantity of his Light. The Shadow of the Earth
likewife terminates before it reaches any of
the fuperior Planets, fo that they are never
eclipfed by us; and the Earth, when fhe is in
Conjunction with the Sun, only appears like
a black Spot upon his Surface.

But *Jupiter* and his Moons mutually
eclipfe each other, as our Earth and Moon
do; as alfo doth *Saturn* and his Moons.
The Satellits of *Jupiter* become twice hid
from us, in one Circulation round ♃, *viz.*
once behind the Body of *Jupiter*, *i. e* when
they are in the right Line joining the Centers
of the Earth and ♃; and again they become
invifible when they enter the Shadow of
Jupiter, which happens when they are at
their full as feen from ♃, at which times they
alfo fuffer Eclipfes : which Eclipfes happen to
them, after the fame manner as they do to
our Moon, by the interpofition of the Earth
betwixt her and the Sun.

Let S be the Sun, A B T the Earth's Orbit; and C ♃ D an Arch of *Jupiter's* Or- *Fig. 7.*
bit, in which let *Jupiter* be in the Point ♃;
and

and let C F D H be the Orbit of one of *Jupiter*'s Satellits, which we'll here suppose to be the fartheft from him. Thefe Satellits while they move thro' the inferior Parts of their Orbits, *viz.* from D thro' H, I to C, feem from the Earth and the Sun to have a retrograde Motion; but when they are in the fuperior part of their Orbit, they are then feen to move from *Weft* to *Eaft* according to their true Motion. Now while they defcribe the fuperior part of their Orbits, they will be twice hid from the Earth, once in the Shadow of ♃, and once behind his Body. If *Jupiter* be more Wefterly than the Sun, that is, when the Earth is in A, they'll be firft hid in the Shadow F, and afterwards behind the Body of ♃ in G: But when the Earth is in B then they are firft hid behind ♃'s Body in E, and afterwards fall into the Shadow F. While thefe Satellits defcribe the inferior parts of their Orbits, they only once difappear, which may be either in I or H, according to the Pofition of the Earth, in which places they cannot be diftinguifhed from the Body of *Jupiter*.

When the Satellits feen from ♃ are in Conjunction with the Sun, their Shadows will then fall upon ♃, and fome part of his Body be involved in Darknefs, to which part the Sun will be totally eclipfed.

By obferving the Eclipfes of *Jupiter*'s Satelliits, it was firft difcover'd that Light is

not

Plate 5

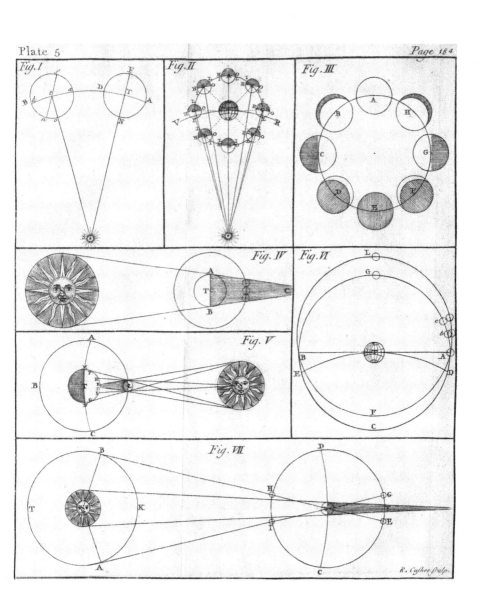

Fig. I

Fig. II

Fig. III

Fig. IV

Fig. VI

Fig. V

Fig. VII

R. Cushee Sculp.

not propagated inftantaneoufly, tho' it moves with an incredible fwiftnefs. For if Light came to us in an Inftant, an Obferver in T would fee an Eclipfe of one of thefe Satellits, at the fame time that another in K would. But it has been found by Obfervations, that when the Earth is in K at her neareft diftance from *Jupiter*, thefe Eclipfes happen much fooner than when fhe is in T. Now having the difference of Time betwixt thefe Appearances in K and T, we may find the Length of Time, the Light takes in paffing from K to T, which Space is equal to the Diameter of the Earth's Annual Orb. By thefe kind of Obfervations, it has been found that Light reaches from the Sun to us, in the fpace of eleven Minutes of Time, which is at leaft at the rate of 100,000 Miles in a Second.

F I N I S.

A N
I N D E X
O F   T H E
Aſtronomical TERMS
Made Uſe of in this BOOK.

Aſcen-

Equi-

Printed in the United States
By Bookmasters